back to nature

back to nature

How to love life – and save it

CHRIS PACKHAM

and

MEGAN MCCUBBIN

First published in Great Britain in 2020 by Two Roads
An Imprint of John Murray Press
An Hachette UK company

1

The lyrics from 'Emergency' by 999 on p278 are
reproduced with permission from UA Records
Illustrations by Sterre Verbokkem

A CIP catalogue record for this title is available from the British Library

Hardback ISBN 9781529350395
Trade Paperback ISBN 9781529350401
eBook ISBN 9781529350418

Typeset in Simoncini Garamond by Hewer Text UK Ltd, Edinburgh
Printed and bound in Great Britain by Clays Ltd, Elcograf S.p.A.

John Murray policy is to use papers that are natural, renewable and
recyclable products and made from wood grown in sustainable forests.
The logging and manufacturing processes are expected to conform
to the environmental regulations of the country of origin.

Two Roads
Carmelite House
50 Victoria Embankment
London EC4Y 0DZ

www.tworoadsbooks.com

This book is dedicated to the wildcat, hazel dormouse, water vole, turtle dove, stag beetle, hedgehog and all other species facing extinction.

CONTENTS

INTRODUCTION

It started with a song thrush and a patch of celandines. Melody incarnate and sparkling stars of gold. The rich dripping of a cascade of notes raw from nature and the bright brilliance of petals shouting 'spring!' That thrush started before dawn in February and was still drenching us in ecstasy in the violet shade of July's evenings; those flowers splashed resolute cheer for just a few weeks but lit a fuse that fuelled a cascade of simple beauty from everyday things that carried us through lockdown and beyond. Our spring of surprises germinated into a summer of love for nature. For some of us, it was with a richness we'd never dared dream would come again, for others, it exploded like a firework as they listened to and knelt down to say hello to a world they had walked past or trodden over for years. For us the joy was in meeting old friends with familiar faces, but in truth this was utterly usurped as we witnessed the palpable rapture of a world waking up to wildlife. Every day, a surfeit of social media excitement burst from the balconies, gardens, parks and green places where isolating people found they were not in fact isolated at all – they were connected to a world of riches they had never before imagined.

That morning fidgeting poodles had me up and out early and when I got back home I thought, 'look at those little flowers beneath my horse chestnut tree' – all clamouring for the attention of the bumblebee queens loping around the garden in a splash of thin sunlight. They struck me as so instantly uplifting that I wanted to share them with people waking up to a new kind of fear, so I sat down with my mobile and beamed them to that little bit of the waking world I could connect with.

One of the viewers was Fabian Harrison. I had met him in the soggy crowd at the Walk for Wildlife in 2018, where he'd come up to me and said – well, to paraphrase it – 'you are not maximising your social media potential because ...' and then a whole lot of stuff that I didn't understand. He looked about fifteen – he was actually nineteen. I like working with young people; they have fewer boundaries, they are less scared of risk, and ultimately Fabian had probably already forgotten more than I would ever know about managing the technical side of social media. There was a simple caveat for his help: my passwords. Umm, given the grief that I get from certain antagonistic fraternities that might have been an unsurmountable obstacle, but not for me. I might be nearly sixty, but I still take risks too. And that one has paid off for us all. The bloke is a unique wizard. He's a birder, a naturalist, an informed and opinionated conservationist, he likes a bit more sleep than me, but is a pedantic perfectionist after my own psyche and he hates the thought of not maximising anything and will not entertain notions of failure. He works bloody hard too, so at this point in time I claim the crown of being his biggest fan. He will soon have many more.

Anyway, the very next morning my two-minute-twenty-second Twitter video was live on that platform as well as Facebook and

YouTube. People seemed to like it, so within a week the 'Self-Isolating Bird Club' had formed with Fabian Harrison as our budding producer and my marvellous PA Cate Crocker running all the burgeoning behind-the-scenes organisation.

So for a few mornings I raved about a series of wildflowers that were bursting out of my patch on a hand-held mobile phone, saying whatever came out of, or off the top of, my head. Soon, as we retreated into our homes and gardens, if we were lucky enough to have one, more and more people began to tune in at 9 a.m. for the little bit of love I had for these humble little things. But within a week there was a problem-ette; I had an outstanding and necessary commitment to *The One Show* and Secret World Wildlife Rescue to make a film about their impending plight under the forthcoming restrictions. I needed a 'stand-in' – cue Megan, no stand-in, but a competent replacement. She is my stepdaughter and had come to 'lock down' with me at the farmhouse. So she did the broadcast pretty well, and then started getting up to join me to do them all. She was no rookie of course; she's a zoologist and had been doing a few of her own presenting projects already, and having two of us made everything more of a conversation instead of a didactic Chris show/rant. Her enthusiasm and youth hit a mark, and her keen nose for fresh science even woke me up too!

Our audience grew rapidly – there was joy in that there nature stuff. Within a week Fabian had invented a new way of using our basic technology and our ambitions expanded and we thought ... maybe we could make a sort of 'programme'? So Cate began managing our contributors and our daily broadcast spread to a busy hour. We reached out to our colleagues

– presenters like Michaela Strachan, Iolo Williams and Gillian Burke mucked in along with filmmakers like Luke Massey, Robin Smith and Matt Moran. Some of our very best wildlife photographers contributed too – Andy Rouse, Paul Goldstein and Matt Doogue all gave generously of their time and the inimitable Gary Moore, wildlife sound recordist extraordinaire got up early for us – but also, critically, lots of other exceptional but less well-known talent willingly helped make it work. Most importantly, and true to my mission, plenty of raw, brilliant young filmmakers, presenters, conservationists and campaigners appeared and they have all been especially inspirational. All together we rambled through an hour of science, culture, photography, films and news reports on Twitter, Facebook and YouTube. We had quizzes, guests and poodle puppies in slow motion – yes, the antics of my poodles, Sid and Nancy, were as popular as Megan sniffing badger poo or the queen hornet on the end of my finger.

By the time we had to take an enforced break for *Springwatch*, we had broadcast 34 hours of our low-tech, unscripted and unrehearsed but passion-packed programming to an audience of eight million. We had regular viewers from Australia to Azerbaijan, aged from four to 94: birders, bug nuts, otter fiends ... the entire breadth of biological interests. But best of all we had hundreds of thousands of newbies frenetically posting comments and photos of the myriad plants and animals they were discovering on their doorsteps. Their excitement was palpable, their joy was infectious, and it spread out and around the entire country and all over the world.

One of the principal reactions people were sending us – beyond 'wow, look at that!' – was the instantaneous awareness of the mental health benefits of connecting with nature. In a time of unprecedented stress, when many people were thrust into the most difficult periods of their lives, they found that nature was there for them, to reduce anxiety, promote calm and provide comfort. No wonder that when polls were conducted after the 'lockdown' only 9 per cent of people wanted to return to 'normal', citing their reasons as noticeably cleaner air, more wildlife and a stronger sense of community. They had come to realise that the 'old normal' was actually abjectly abnormal, that going back to business as usual would be bad, as it had been 'bad business', certainly when it comes to the environment. With the blame for this pandemic focused upon China's live wild animal markets, with their horrific menagerie of animals smuggled from all over the world, it was plain for all to see that abusing the natural world was exacting a terrible cost to everyone globally. While for the citizens of the UK the Californian and Australian wildfires, the destruction of the rainforest and the melting poles all seemed far away, suddenly a lethal virus, which had possibly jumped from a bat to a pangolin, was potentially in their home ... or in them. This was the unsustainable abuse of our Earth writ large and the full horror of its repercussions still remain unimaginable. Only fools would want to go back to that ... but then fools are who we have in charge.

In June, under the auspices of Wildlife and Countryside Link, the CEOs of the majority of the UK's conservation and environmental NGOs wrote to the chancellor Rishi Sunak and George Eustice, the Secretary of State for the Environment, with a plan. They placed on their table the offer of a rapid economic recovery from the coronavirus crisis that would provide a more resilient society by restoring natural capital, the natural wealth that underpins our wellbeing and a productive economy. The proposals highlighted the need for investment in our natural environment – if undertaken they would quickly create jobs, save billions of pounds for the NHS, level up health and livelihoods and protect the economy against future natural disasters. Of course this green recovery would need to be supported by a rapid investment in a low-carbon economy to address the declared climate and environment emergency. They included a list of ready-to-go projects that would generate employment immediately, and asked for £315 million – not a lot of money to create 200,000 hectares of priority terrestrial and marine habitat in a new Nature Recovery Network, to reverse declines in biodiversity, to create 10,000 jobs, support rural and urban economies countrywide, lock away millions of tonnes of carbon in pursuit of the UK's net zero target, and protect people and businesses from future natural disasters. In response, Rishi and George came up with £40 million, just enough to plug the gap due to the pandemic. Our government have subsequently announced a huge investment in road building and fought me through the courts to ensure that the £100 billion HS2 high-speed rail behemoth continues to smash its way through our ancient woodlands. That is not a 'green new deal', that's 'carrying on as normal'. The old 'normal', the 'normal' that got us all in this mess.

We have spent our spring and summer finding respite and solace in nature. It has been there for us, now we need to be there for it. The forces that are destroying it haven't gone away while we have necessarily been preoccupied with looking after ourselves. But now we all know that more than ever, we need to look after nature to look after ourselves.

In this book Megan is going to excite you with some gems of extraordinary new science – almost unbelievable nuggets that you'll want to share – we are going to celebrate some of conservation's successes, and explore the methods we all have to generate more in our own spaces or communities. But it is also my duty to expose some calamities and crimes that are continuing to thwart our urgent endeavours. I'm giving you the facts, but as ever we don't expect or even want you to agree with all our opinions. What we want you to do is think about them and formulate your own – all we need to make progress is commonality, not complete agreement. We need to change – that most difficult of human challenges – and because some of that change needs to happen very, very quickly, we will meet reluctance and outright resistance. We are not dismayed by this; to us it's a nuisance, nothing more. We are going to make a last stand for nature because we have no choice. We have to be the voice of the persecuted and oppressed. Our job is to translate the notes of the song thrush and reflect the beauty of those celandines until enough of the world loves life – all life – to cherish and protect it.

LOVE LIFE – ALL LIFE

I remember, with exquisite clarity, climbing onto a low wall by my parents' gate when I was four and repeatedly reaching into their neighbours' bush to retrieve a multitude of ladybirds, which, with the benefit of hindsight cruelly and inappropriately, I incarcerated in empty matchboxes. I was fascinated by their variety and entranced by their simple beauty. I admired their symmetry, laughed as their little feet tickled my palms, winced when I licked their yellowy bile that blobbed onto my nails and, best of all, I loved them when they wound their way up my fingers, twirled around, twisted and twitched before peeling back their shiny red wing cases, unfolding their sparkling wings and wafting up into the sky until they vanished.

I loved ladybirds, I was obsessed with ladybirds, I was addicted to ladybirds. Every morning, as soon as I had gobbled up my breakfast, I ran to clamber onto that wall. Next came caterpillars: thin ones, fat ones, spiky ones, hairy ones, all trundling over my hands, curling, uncurling, stopping to wave their beaded heads, pooing little pellets into my palms. Palms that were soon full of tadpoles, cupping minnows and sticklebacks, getting stung by wasps, scratched by beetles and bitten by water boatmen when I squeezed my hands too tight. And as much as I craved the feel of nature, and always wanted to touch it, I also listened to it; the house sparrows chattering on the gutter, the blackbird on the TV aerial, the sounds of those caterpillars munching leaves, the scraping of a violet ground beetle as it spun upside-down on the cement path, the clattering of swifts coming to roost in the loft above my bedroom. I tasted it too, licking the creamy ooze that encircled picked dandelion stalks and the frothy gobs of cuckoo spit that dotted that same ladybird bush, and – infamously – a number of those tadpoles that wriggled in jam jars on my windowsill and then around my tongue. Smells were intoxicating too, some judged bad, others sweet – the bitter twang of a wounded beetle versus the powdery paper scent that I sniffed from beneath the wings of a dead starling I found at the bus stop. I was inextricably drawn to nature; as soon as I could waddle its magnetic attraction pulled me into its myriad beauty, where I spun, entranced and ever hungry for more, more, more. I was born a biophile; the love of life was in my blood or, more accurately, in my genes.

DELICATE AND STRONG

The vast quantities of pesticides used throughout the UK are of course non-selective, killing any and all insects in the area, even the ones that are natural 'pest' controllers themselves. Ladybirds have long been favoured by farmers as they avidly consume aphids and other insects, which damage crops. There are over 40 species of ladybirds that are residents in the UK, though they are in decline mainly due to the presence of the invasive non-native harlequin ladybird, but also from the additional pressures of pesticide use.

We have all picked up a ladybird or two in our time. If you have ever watched a ladybird fly, have you noticed how their wings are much longer than their bodies? The red or orange spotted 'wing cases' on the outside of their bodies are actually modified forewings called the elytra, and they rely on their elongated hindwings to give them flight. The length of the hindwing in comparison to the body is quite disproportionate and, up until 2017, scientists didn't know which mechanisms were used to fold the wings neatly under the elytra. This is because the elytra close after flight but before the wings are taken in, making it impossible to see.

Determined to find out what was going on, researchers at the University of Tokyo fitted ladybirds with an artificial, transparent elytra and filmed them taking off and landing in slow motion. It turns out that the wing frame does not have a joint, which is unusual for structures that can transform and be folded. Instead, the wings can fold up like origami due to the flexibility and elasticity of the veins around the outside of the wing, which allows creases to form. We could potentially learn a lot from this strategy as these little beetles have got the perfect balance for the strength and stability needed for flying, but also enough flexibility and instability for folding without any rigid joints. Engineering at its finest and most delicate.

The term 'biophilia' was first coined by German-born American psychoanalyst Erich Fromm in 1973 when he described it as 'the passionate love of life and of all that is alive'. It encompasses an idea that humans harbour an innate tendency to seek connections with nature and other forms of life. The brilliant American biologist EO Wilson expanded the thesis in his 1984 work *Biophilia* by suggesting that such a tendency to be drawn to nature and to affiliate to it and other forms of life has, in part at least, a genetic basis.

Just as I was sucked into an inexplicable attraction to those simple commonplace organisms that lived around our house and garden, many others manifest a similar appreciation. People love the rich diversity of shapes and colours in life and respond to them creatively. We have also integrated nature into our languages and very notably into our religions; there is a ubiquitous spiritual reverence for animals in cultures all over the world. This spiritual experience and the way nature has been woven into the fabric of our lives arose at a time when we lived in far closer contact with the natural world than we do today. Indeed, there is little doubt that in the West those bonds between us and other life sadly began to unravel in parallel with our technological progress in the industrial and social revolutions of the nineteenth and twentieth centuries. These took us away from nature in a physical sense, as we began to inhabit spaces devoid of as much wildness. Now in the twenty-first century we live in sterile modern homes, where we can't see, hear, touch or taste it, and we drive through it in our air-conditioned cars, disconnected from it. It's in these environments that we spend more and more of our time, to the point now where people have switched from biophila to biophobia, that fear of life that fuels our intolerance and ignorance of the natural world. Of course, some 'nature phobias' are ancient, originating when we were vulnerable to predators and poisonous things, and such fears were actually part and parcel of that connection to our environment – they kept us alive in it. But the abundance of modern phobias and intolerances are unrelated; they are manifestations of a circle of reinforced and repeated ignorance. As I've mused before, why are there so many toxins on supermarket shelves to kill ants? How many people die in ant attacks?

But it's not just a physical separation that is deconstructing our biophilic urges. Culturally too we are disconnecting. The words and names of natural things are disappearing from our lexicon. In 2007, the Oxford University Press released a new edition of its junior dictionary, aimed at children aged seven and older. Words like 'moss', 'conker', 'kingfisher', 'newt', 'blackberry' and 'bluebell' had gone missing. Even 'otter' and 'magpie' failed to make the cut and in their place they had added 'blog', 'chatroom' and 'database'. The publishers haughtily justified these omissions by proclaiming that it was their duty to present words that were in common parlance and that these words had fallen out of use – a small conker some might say, but an insidious reflection of waning biophila. Thankfully in a measured and beautiful way author and wordsmith Robert Macfarlane and artist Jackie Morris retorted with their gorgeous *The Lost Words*, a book that celebrates 20 of these excised nouns with acrostic poems, spells to be read aloud to conjure up everything from goldfinch to wren, from adder to weasel. It has become an acclaimed bestseller – so stick that up your dictionary!

NOT YOUR AVERAGE EGG

Inspecting birds' nests tucked away within the bushes was once a common activity, but today, unless trained and for a scientific purpose, we know we must keep our distance to minimise any disturbance that could cause undue stress to the birds. While we should no longer be peering into their nests, birds often leave behind clues or indications in the form of eggshells.

Eggs are some of the most delicate and beautiful structures of biological architecture. Birds' eggs have the most variety of colours, shapes, sizes, brightness and shines, which is something that has captured the attention of some of the best zoological minds over time. But how are they made and what creates such variation? An egg takes approximately 24 hours to form inside the female. The ovum (or egg cell) is released into the oviduct, where it can be fertilised. At this point it is just the beginnings of the yolk, packed full of protein. The albumen comes next, which is the gelatinous egg white that surrounds the yolk, and this mass then expands with the addition of water, and a stretchy soft membrane forms around it. Particles of calcium carbonate that create the shell then come from specialised cells lining in the shell gland, and finally comes pigmentation. The base coat is followed by the species-specific spots or blemishes that are genetically programmed to fire like paint guns onto the shells' surface. All colouring comes from just two organic compounds: firstly there is protoporphyrin, which produces the red-brown tones, and secondly biliverdin, which is responsible for the production of red and green. The colouration of birds' eggs only comes in the last couple of hours of egg formation, which makes it really difficult to study exactly *how* the colours make it on there.

However, there are many theories as to why eggs are patterned the way that they are. In some cases, like that of ground-nesting birds such as lapwings, eggs are coloured for camouflage against predation. Others, like the wren, have brightly coloured blue eggs to help the parental birds locate them when nestled in a well-camouflaged nest. Some other ideas are that colouration helps to protect the embryo from harmful UV light or that the different colours may even have varying antimicrobial properties. All these affect pigmentation on a localised level and there's been one breakthrough that helps fit another piece of the puzzle – scientists investigated the eggs of 634 species across the globe and found that eggs in colder places were generally a dark brown, while there was a lot more variety in the tropic and temperate regions with blue-green, white and brown-red eggs. The darker the eggs, the more heat-absorbance is possible, which is a big advantage in colder regions. Eggs in warmer environments do not need to absorb so much heat and so they would have evolved in parallel with the selection pressures mentioned above, like predation risk or antimicrobial properties, and are more varied in their colouration. How this will evolve – and whether it will evolve in time – with rising temperatures and climate change remains to be seen.

Are we being pedantic, or does what we now describe as 'Nature Deficit Disorder' (NDD) – not actually a medically recognised mental health condition, but an increasingly respected and cited reason for children's, and adults', disconnection from nature – really matter? In a nutshell (if they ever get to pick one up, because yes, nutshells themselves are getting rarer, most nuts now being presented to us in plastic packets): yes.

In the case of children NDD develops mostly as a result of parents keeping them indoors, ostensibly to protect them from danger, fearing threats to their safety and the risk of injury. Of course in parallel there has been a decline in the accessibility of natural surroundings in many communities, and even parks and nature reserves often have restricted access and plenty of 'do not do this, that, and the other' signs. We conservationists also now seemingly discourage young people from physically connecting with nature with our 'look but don't touch' mantras, presumably manifest because we fear that little fingers may be clumsy and squash the frog, flower or fungi. And even when children are taken by teachers into natural spaces they have to wear high-viz jackets and constantly get doused with hand sanitiser as soon as they touch anything vaguely 'dirty'; less than subconscious signs that they are entering dangerous and unhygienic environments. One of the worst mornings of my life was spent rushing to an urban wildlife site in London for a photo call with the then environment minister, arriving late (trains), and running up to a pond where dipping was taking place with all the children wearing bright yellow rubber gloves. I refused to be in any photos with a gloved child; the organisers hated me, but the practice was almost immediately rescinded, and the youngsters were able to feel newts again ...

Coupled with the increased fear of 'stranger danger', a phenomenon that is constantly and furiously fuelled by the media, this I thought about it a atmosphere is keeping children indoors rather than outdoors exploring. Ironically, they are inevitably spending this time watching TV or they are online, gaming or using social media, all things that may expose them to far more horrors and far more dangers than they would encounter in natural spaces. The average American child now spends 44 hours a week engaged with electronic media and a 2018 poll revealed that in the UK young children spend twice as long looking at screens as they do playing outside. By the time they reach seven children will have been looking at screens for the equivalent of 456 days, an average of four hours a day. In comparison they will have spent just 182 days, or an average of just over an hour and a half a day, playing outdoors. I hate to court controversy again, and I know opportunities are restricted, access to safe outdoor environments can be impossible to find and pressures are tight for many, but perhaps we could make a case for parents, grandparents, guardians and carers, teachers, adults in general, having too much control over what children are doing with their lives. Given a real choice I think a lot more kids would happily be 'feral'.

Whether it's medically recognised or not, the repercussions of NDD are now coming home to roost. Without access to any nature, people, young and old, cannot develop or maintain that innate biophilic urge and therefore for them nature loses its meaning. They lose respect for the natural world and display a decreased appreciation for the diversity of life-forms that support their survival. Is it any wonder that they struggle to engage with any resistance to environmental destruction and the escalating rate of species extinction? But there is a flip side to this, in that we humans don't know what we've got till it's gone, then we always seem to want what we can't have. When we wake up to the absence of nature, we begin to crave it – in short the grass is always going to be greener if you haven't got any grass.

I think a key part of our incredible connection or reconnection to nature under lockdown was due to our biophilic urges being unleashed. When we were extracted from our rigid schedules and removed from those principally nature-denuded environments of travel, work or school, we were able to cast aside some of our technologies and coincidentally spend more time outdoors, and those few hours a day lit an ancient innate fuse that woke us up to that long-hidden desire and need to know nature for our own good, for our survival. And once exposed it proved irrepressible and contagious, because by connecting with nature we were actually becoming more human again, more complete and more healthy organisms. No wonder then that we got so excited! And no wonder that so few of us – 9 per cent, remember – wanted to get 'back to normal'. Under a dark cloud of fear and confusion people all over the world found solace and respite in nature; it improved the quality of their lives and their physical and mental health. And, given the stresses we were all under, it's not surprising that the latter aspect drew considerable attention.

Regardless of the extent to which people feel or even understand their biophilic potential, research has indicated that spending time in green spaces is beneficial for mental health

– even spending just two hours a week interacting with nature generates a greater satisfaction with life than experienced by those who spend less time in natural environments. Many of us had known this or had come to know it long before this research. When I was a child and a troubled teenager and twenty-something, I thought that I was spending most of my time 'with my head in a bush', as my mother chose to phrase it, because it was there that I would find those things which fascinated me most. But with the benefit of hindsight I was also there to alleviate my anxieties and to be calm away from the conflicts that cramped my life. I felt better in the quiet, more comfortable alone with no need to compare myself or my behaviour with that of my peers. I liked the sensory experiences, which were so much greater than those 'indoors', but most of all I just felt more in-tune with the world, with my own time and space. It's no exaggeration to say that although I had no understanding of my mental health issues – I was just dogged by them – the treatment my access to nature unwittingly offered saved my life. I spent a lot of time crying in dark woods and rainy fields, but nature and wildlife soaked up those tears and sometimes even put a smile on my face.

I think my handicap in understanding and accepting that I was actually self-medicating was my unremitting refusal to acknowledge anything metaphysical in myself or my world. If it wasn't tangible, however tiny or remote, if it couldn't be described and quantified then to me it didn't exist. Later as an adult (maybe in my forties; it took a long time for me to get there) I could no longer refute the existence of my feelings and how they were positively influenced by just being out in the woods, principally with my dogs, Itchy and Scratchy. This just meant that I had to go looking for those real empirical proofs that the two were connected. That meant science – scientific evidence, peer-reviewed studies – and at that time there wasn't a lot of it. Things have changed dramatically now.

Let's go first to the work of Professor Miles Richardson and his 'Nature Connectedness Research Group' at the University of Derby. In an overcrowded urban world where mental health disorders affect 30 per cent of the global population, Richardson felt that simple nature-based solutions were often overlooked. His 2019 paper in the *International Journal of Environmental Research and Public Health* shows how an increased connection with nature can establish clinically significant improvements in quality of life for people living with a mental health difficulty, and in fact for all adults. All through simply registering the good things in nature in everyday life. He and his fellow researchers found that people who spent less time outdoors in the previous year improved more when they experienced nature connectedness. Further, those who had lower baseline nature-connectedness scores improved even more. As a consequence this suggests that specific targeting of those who spend little time outside with a simple engagement strategy would improve their mental health. As it turns out, the application of this therapy came to many in need in the form of the 'lockdown'; what seemed initially to be a serious restriction to our lives and freedoms in fact for some became a real opportunity to employ a mental health remedy, provided they had access to natural environments. Also this lockdown re-connection will probably pay more dividends down the line as the study appeared to reveal that those who had spent more time outdoors as a child showed a greater improvement in their nature-connectedness scores. Whether childhood exposure to nature is important for nature connectedness as an adult has not been empirically established, so this could hark back to that proposed genetical link to biophilia in the form of a 'latent nature connectedness'. Of course we know that any childhood connection all but disappears during adolescence for many young people, but it might be that an early connection with nature can be reignited and that this results in subsequent wellbeing benefits – again something many people seemed to report during the lockdown euphoria.

But how is it actually working? What are the actual mechanisms for mental change? Richardson explains that 'positive emotions are a reflection of what is known as "hedonic wellbeing", which relates to feeling good, the pleasantness of our experiences and the extent to which our desires are fulfilled'. What is also important for our wellbeing is our ability to function well psychologically. This is referred to as 'eudaimonic wellbeing', and includes factors such as autonomy, self-acceptance, having a meaning and purpose in life, and recognising personal growth.

We know that eudaimonic wellbeing is related to, but distinct from, hedonic wellbeing, that each are important and that people with high levels of both types of wellbeing are considered to be flourishing. Although Richardson's work continues, one can suppose that a balance of these two emotional states must be stimulated by the experience of nature connectedness, and that this should lead towards changes in the way we prescribe treatments and therapies to address mental health. And of course it's a win-win, because as we tune our senses, appreciate its beauty, celebrate its meaning and respond emotionally, we are vigorously activating our compassion for nature, and we will be kinder to it.

Before we leave the topic of mental health I'm worried that many of you won't be readers of the *International Journal of Environmental Research and Public Health*, so I must draw your attention to my current go-to guide to natural wellbeing, *The Wild Remedy* by Emma Mitchell, who is also a wonderful contributor to the Self-Isolating Bird Club broadcasts. The book takes the form of a charming hand-illustrated diary with drawings, paintings and photographs, and charts Emma's highs and deep lows in terms of her own mental health through a year of her rambles as they radiate from her home in Cambridgeshire. Emma has suffered with depression, or the 'grey slug' as she names it in parallel to Winston Churchill's 'black dog'. Churchill's psychological spectre snarled at him for most of his life; Emma has been 'slimed' by her negative moods for more than twenty years. It's a candid account, she doesn't hold back or hide her issues, but like her it's also a gentle, witty, reasonable and ultimately persuasive chronicle of how our minds and bodies can respond positively when we get out into greener spaces. But I think crucially Emma is also a scientist, and she presents a catalogue of remarkable recently published research that will confirm to any sceptics that engaging with nature offers a very real and measured cure, one that she uses to 'punch and trample' her 'great grey mollusc'. You can follow her progress and admire her stunning arts and crafts on social media where she is 'Silverpebble'.

SWEET DREAMS

Practice makes perfect, as they say. This is a phase we hear a lot as we go through life learning new skills, whether that be walking, drawing or playing the flute in preparation for an audition at the Royal Philharmonic Orchestra. Our days are packed full of information, some of which is useful and some of which is not, so when our bodies shut down to sleep our brains go into overdrive replaying the day's events, filtering out which memories and new skills are worth hanging onto. The dreams we have essentially help with memory consolidation, but dreaming isn't solely limited to humans and evidence of dreams can be observed in many different species, from cats to cuttlefish. Of course, it depends on your definition of dreaming, but scientists are able to measure brain activity in animals as they sleep and find that many species, similarly to us, will have rapid eye movement (REM) cycles. This is the point of sleep where vivid dreams generally occur.

For the last 20 years or so, we have known that sleeping birds have neurons firing around their brains in a complex pattern similar to that observed when they are awake and singing. This suggested that birds are able to subconsciously practise and tune their songs while sleeping – a skill that is particularly useful for young birds, who develop their singing abilities by listening to, remembering and mimicking that of a parent or related (conspecific) individual. It's simply fascinating that they replay these songs in their minds as they dream to improve their skills. But when investigating this further, scientists found that zebra finches also flexed their vocal muscles when asleep as if they were actually singing. The results suggest that birds could be practising variations of their song, which helps to fine-tune their notes to make sure their song is perfectly composed. The vocal muscle activity was so strong that if enough airflow was present, they'd actually be singing as they sleep. What a beautiful thing that is.

As the lockdown intensified and increasing numbers of people were confined to their homes the Self Isolating Bird Club grew and grew, and so did the BBC's realisation that 2020's *Springwatch* would be both an enormous challenge and a very necessary tonic for the ongoing crisis. Behind the scenes the always hard-working team worked even harder to shape a multitude of methods to make the shows, whichever scenario was playing out in terms of restrictions at the end of May. In the end the presenters stayed home: Megan and me in the New Forest, Iolo in Wales and Gillian in Cornwall, and very sadly Michaela at her home in South Africa, so she couldn't appear this year. We were joined by two camera operators and two engineers: we all diligently observed the government's social distancing guidelines and prepared to put the programme on air. The technical challenge was massive. The team, who normally assemble in several large trucks and portacabins, were scattered all over the UK, working from their offices, lounges and bedrooms. Meetings were done by Zoom, we watched the wildlife footage through Dropbox, and we sent a lot of emails and text messages. On Tuesday 26 May someone somewhere pressed a button and somehow our titles ran on televisions across the nation.

It was sunny; spring had been warm and wonderful for wildlife, but terrifying for humans. I knew more than ever that when the theme music stopped and a voice in my ear said 'cue Chris' I had to hit not just the mark but the right mood, the mood that would be respectful of a population in the grip of a calamity, but which would also promise them a breather, some natural escapism and a reprieve. I never normally think too much about what I'm going to say words-wise – I'm a horror to rehearse with, I'm a 'from the hip, in the moment, use the force' sort of presenter; I like making up stories as I go along, not following a narrative written by committee. But that night at about five to eight I thought about it a bit ...

I'll jump to the end of the *Springwatch* 2020 story. We scored a hit; everyone dug deep and delivered, we reported some amazing wildlife, using our mini-cameras we stuck the nation's noses into the intimate lives of some birds and, given our not insignificant constraints, suffered only one minor fuzzy moment that lasted a few seconds. I can tell you it was 'seat of the pants' stuff much of the time for everyone involved, but then our team are the best and, as I often like to tease our viewers, 'that's what you pay you licence fee for': a unique collective of the world's best technical talent and natural history programming expertise, all for 43 pence a day. Independent and impartial too, and along with the NHS, RSPCA and the *Big Issue*, one of the best brands left in Britain. At a time when they are frequently falling, we got our highest viewing figures for years and an exceptional audience response – a very clear indication that we love our wildlife and turn to it when we are in trouble. The nation didn't just tune into *Springwatch*, it tuned into nature as never before. A nation of shopkeepers became a nation of biophiliacs.

And of course if we love something we will want to cherish and protect it. But sadly for all our efforts it gives me no joy to reinforce the obvious: to date, and overall, conservation has failed to halt the widespread declines in our wildlife and its habitats. My optimism has increased however, given this recent upsurge in nature connectedness, as I'm sure that – aside from the quality of life enhancements and natural therapies – more people will recognise our species' practical dependence on nature; what I call 'Ecosystem Services Earth'. This global resources company – owned by none, abused by many – provides us with clean air and water, functional soils in which to grow crops, trees for timber and fuels, productive ocean ecosystems from which we harvest fish, and when we are sensible waves, wind and sunlight to give us power. We have all just been reminded that when nature is healthy our species should be healthy; when it is not, we are not. And, in combination, an awareness of this need to coexist harmoniously – rather than dominate destructively – with our now increased ability to see nature as a source of inspiration and peace, and something we can form meaningful emotional connections with, has to be a turning point. But in order to make it we face a fundamental challenge ... we must change our minds.

Too big, too far? Changing mindsets

As you read this book, Chris and I ask you to keep an open mind. You'll notice a common theme will emerge as the chapters go on, whereby we loosely discuss our human reluctance to change our own mindsets and behaviours, which is often one of the most challenging of steps to overcome. It is a huge barrier when it comes to conservation and environmental protection. I have always assumed that it is because we are somewhat stubborn, and feel that the idea of changing our minds or actions means admitting and facing up to the fact that we have been bad or are in some ways wrong. I can't help but wonder about some fundamental questions: why aren't the scientific facts enough to change our opinions and practices? And why do we react to these facts as if they are a personal attack on us as individuals? So, before we get into the nitty-gritty of how we can cohesively and productively put nature back together again, let's talk about how the brain and the psychology behind these responses works.

Each new day our brain makes thousands of snap decisions – some sources suggest we make up to 35,000 of them daily, and some hold more weight than others. For example, there are the obvious decisions like when to get up, what to wear and what you decide to eat, but in among these bigger decisions there are smaller choices we make almost instantly – and in some cases arguably subconsciously – like swerving to avoid a pothole, scratching your arm or making the choice to continue reading (and thanks very much for that by the way!). But, of course, the type of decisions we are talking about in this book are the conscious and considered choices to change an ingrained and sometimes deep-rooted opinion that may need us to alter a particular behaviour. The region of the brain that is responsible for that kind of decision-making is the prefrontal cortex, which is also the area for planning, personality expression and social behaviour moderation. And the topic that we display most reluctance to change our thoughts on is politics – hardly any surprise there.

One possible explanation is that we confuse our partisan (a.k.a. political) identities with our personal identities and, as such, we take any alternative views or facts as an insult. These insults, which may be entirely scientific and not personal in any way, can feel like an attack and, as our brain is hardwired to defend itself, we can sometimes respond in a defensive and irrational manner. For example, Albert Einstein is renowned as one of the most important physicists in history: we are taught about his theory of relativity in schools, he shaped the way we understand the universe and built the foundation on which many other discoveries have been made. Yet research shows that people are more easily convinced to change their minds to believe that Einstein was not a good physicist than they are to alter their own opinions on immigration and the death penalty. This inflexibility is not necessarily due to intelligence level, but it's just that we are much more willing to be flexible on non-political matters. And unfortunately for us, at the moment, many of the environmental catastrophes we are facing are very closely linked to politics.

For decades now, the global scientific community has been warning us about the current and potential future impacts of a changing climate propelled by anthropogenic activity. We have been told time and time and time and time again that using fossil fuels accelerates global warming, pumps out pollution that reduces air and water quality and destroys vital habitats and wildlife. Yet here we are in 2020, still drilling and mining to collect it before pumping it back out again as greenhouse gasses. It's not that these messages are falling on deaf ears because we are certainly hearing them – they are loud and clear – it's just that we are not willing to make the decision to jump before we are pushed. Why is that? Humans also have an optimism bias, which means that we tell ourselves it'll be sunny tomorrow even when the weather forecast predicts thunder and lightning and the news has issued a serious weather warning. Okay, maybe it's not quite as dramatic as that, but you get what I mean; we prefer to look on the brighter side.

Neuroscientists and psychologists have also used imaging technologies, namely functional magnetic resonance imaging, that enable them to see how the brain makes choices. As a result it is slowly becoming clearer why we behave so foolishly on issues concerning the climate. According to the studies, the brain simply cannot commute or rationalise climate change, because we just aren't programmed to respond quickly to big-scale slow-moving threats. Professor Daniel Gilbert of Harvard University said 'our brain is essentially a get-out-of-the-way machine, that's why we can duck a baseball in milliseconds. While we have come to dominate the planet because of such traits, threats that develop over decades rather than seconds circumvent the brain's alarm system. Many environmentalists say climate change is happening too fast. No, it's happening too slowly. It's not happening nearly quickly enough to get our attention.'

This is particularly interesting when you think about Greta Thunberg and the Fridays For Future youth movement, which has gained serious momentum since 2018. For that generation, and even more so for that of their future children, the impacts of the climate crisis are much more immediate than they are for their parents or grandparents. Perhaps teenagers' and young adults' brains are processing and responding to the impending crisis much more effectively than that of older generations simply because it will have more effect on them? There's an interesting psychological study to be done somewhere here!

Elke Weber, a professor of management and psychology at Columbia University, said 'self-control is a huge issue for people, whether it's what we're eating or saving for our retirement. There's a two-year-old in the back of our minds that's still there that we've learned to overrule that wants to have their one marshmallow now rather than wait for two marshmallows later. Very few people on this planet want to destroy Planet Earth. It's just that our other agendas get in the way of things that might have a longer time horizon'. It's sad, but it does make sense. However, there is something that does appear to work – our brains are hardwired to respond to rewards for good behaviour (we've all done that strut home from school after being awarded a gold star for sitting nicely in assembly or completing all your spellings – right?). If we have an incentive or a metric, something visual to work towards, whether that's gold star stickers, scoring points on video games or watching as your carbon footprint falls down a notch or two, then we tend to maintain our attention for longer as we are focused on a goal we can achieve right now.

I do hope this book challenges your opinions and thoughts in the same way that it challenged ours as we were writing it. I have found new science and read new points of view that I have never considered before. It's been eye-opening. We talk about a lot of difficult issues, but there are also lots of practical solutions to think about. If we are able to give ourselves goals to achieve today, such as taking the bus, eating less meat or joining a protest, then by tomorrow we will be in a better place and ready to tackle what is coming our way ...

A HANDFUL OF DIRT AND A DANDELION

It's a hard watch. Like so many clips we see on social media these days, the images playing out on my laptop are of unimaginable suffering. Many people say 'I just can't look' or 'I don't need to see it', but I believe that we owe it to these poor animals to look because if our views of their mistreatment prompt us to take action, then they won't have died entirely in vain. Plus secondarily, getting this footage often comes with a terrifying risk and psychological cost for those brave undercover filmmakers who go out and collect them. We owe it to them, too. The footage is blurry, but there are mounds of giant bats baked onto sticks, huge beautifully patterned pythons lying in tangled coils on tiled tables, tiny monkeys cowering in cramped cages and, through the bustling crowd of shoppers, you can see dirty crates overflowing with writhing piles of eels, crayfish, ducklings and kittens, all desperately clamouring to escape.

The phone camera then pans onto piles of cooked rats or squirrels, rows of tortoises, terrapins, all lined up on a floor awash with blood, and finally, those strangely scaled bodies of pangolins, the most trafficked wild animal on Earth. This netherworld of animals stolen from all over the globe is my introduction to the nightmarish live animal markets of the Far East, the likely source of the virus that is confusing, terrifying and killing people far away from its evil source. It's a gruesome start to the day and when it's over I close my Mac, trudge across the room to the door and out into my garden. In just one step I go from hell into heaven.

Life. Everywhere. All busy existing in as much harmony as it can find. Flowers and trees, birds and bees, all swirling in a dizzying variety, the sights, the sounds, the smells all so rich in wonder. I breathe and taste spring, that wet grassy urgency of freshness, clean and new. And after that moment during which my senses were swamped, I begin to find the individual parts of this place; the stunning flower candles on the horse chestnut, the sharp song of the robin and cooing of the woodpigeon, the drone of the fat fluffy bees, the twinkle of the wood anemones that constellate in the shade of the hawthorns, all dressed for a wedding, billowing and white, fringing this patch of earth that I have helped make a tiny piece of paradise. My garden could never have spawned the coronavirus, but it will be my salvation from it.

And not just me, millions of others too. Reports suggest that as many as 27 million people in the UK are active gardeners, a significant number of the 66.65 million that currently live here. The gardens where they garden are on average between 14 and 42 square metres in size, depending on the type of housing, and our front gardens are significantly smaller than our back. Some 87 per cent of our properties have a garden, so if we extrapolate from the total of 26,159,440 households in the UK, this equates to 22,738,563 houses having a garden. If we then multiply this number by the respective averages for the various garden areas, we get a total for the UK's garden cover of 432,964 hectares, which is – wait for it – one-fifth the size of Wales. More importantly for us, this adds up to be a greater area than is covered by the Norfolk Broads, Exmoor, Dartmoor and Lake District National Parks put together. So as much as we may have some serious problems with land ownership, acquisition and access in this country (more on this later), the garden owners of the UK have a lot of potential at their green fingertips to do good for wildlife and themselves. Even within our cities, 25 per cent of their area is composed of its private gardens. And we spend money on them too; collectively per year around £7.5 billion on gardening stuff, £2.4 billion on the services of gardeners and individually around £150 on our own gardens.

So that paints a bucolic picture of picturesque urban, suburban and rural floral borders, trellises, greenhouses, vegetable patches, planters, patios and primroses, but sadly all is not what it should be in our modern world. The flipside of our garden life is revealed by data collected this spring by the Office of National Statistics, which showed that 12 per cent of households in Great Britain had no access to a private or shared garden during the coronavirus lockdown. This figure is worse in London where it rose to 21 per cent, the highest percentage of any region of the UK. The next highest was Scotland at 13 per cent. All of a sudden, I hope that many of us are feeling a lot luckier. But it gets worse, because a Natural England survey has uncovered that in England Black people are nearly four times as likely as white people to have no access to outdoor space at home, whether it be a private or shared garden, a patio or even a balcony. So if you are Black and live in London you are a lot less likely to have been able to just step outside and start to relieve your despair. And despair, worry, stress, fear and confusion were part of everyone's 2020 spring.

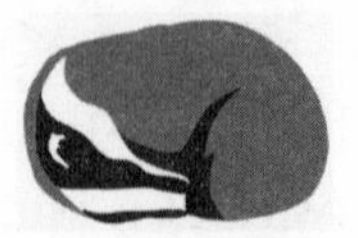

The lockdown proper was imposed on the evening of 23 March and from that point the vast majority of us were confined to the boundaries of our properties – isolated, only venturing out for essential items and unable to connect with any separated friends and family. But this separation and isolation was ironically what many people now had in common, and it soon began to connect us in an unexpected and then overwhelming way. After years of being too busy we got to know the neighbours over the fence, rekindled old friendships online and made calls to vulnerable family members a priority. Megan and I 'locked down' on the farm I rent in the New Forest; we'd go out once on Saturday to a small local shop to get what we, my father and her grandparents needed, then drive to their houses and stand or sit outside for a few hours. For the other six days a week, for more than eight weeks up until we started *Springwatch*, we didn't go out, or see anyone. Now, I'd perversely been in training for this for years; I will regularly self-lock down on my own, in the woods with the poodles, and refuse to go out unless it's to work. But that's me; I have what was known as Asperger's, a form of autism, and feeling comfortable in my own company is part of a personal management strategy that gets me through ... But Megs is 25, sociable and has friends in a way that I don't, so she was soon struggling to try to normalise this very abnormal period. Online quizzes blossomed as the world woke up to Zoom, ever more ambitious cookery kept her in the kitchen experimenting with complex recipes and home-baking. Most of the concoctions were great, but the smoke alarms were tested and the foxes were never short of table waste. Our days were busy with the Self-Isolating Bird Club broadcasts and each evening ended with #PunkRockMidnight as I trawled through my vinyl for seventies classics and posted my celebrations on Twitter at the witching hour. In between we re-watched all of *Twin Peaks* and all our favourite sci-fi movies, read a lot and spent any spare daytime hours taking photographs, mainly in the garden.

For many people in the UK and Europe our fine spring weather was a saviour, as we could get out into those gardens and green spaces for our daily hour of exercise without wellies, fleeces and raincoats. It was the sunniest spring since records began in 1929, with over 626 hours of pure brilliant sunshine, the previous record having been set way back in 1948. It was also the eighth warmest spring on record and the fifth driest. In fact, it was the driest May ever in England and some counties in the north-east and in eastern Scotland recorded their driest spring in a rainfall series that started in 1862. Environmentally these are ill omens, but imagine if it had been the darkest, coldest and wettest ever – that would have been cruel and undoubtedly disastrous as we have needed our gardens and greenery like never before.

Yes, our gardens and local green spaces became a refuge, and post-lockdown nearly everyone (97 per cent) who responded to a Wildlife Trust survey said that throughout the course of the restrictions nature had been important for relieving stress, and as a result 85 per cent said they are more likely to now help wildlife where they live. Many were 'living' in their gardens, that safe space that had allowed them to reconnect with the wild. Indoors, the news was all bad – a bad thing doing bad things to people, or bad people doing bad things because they couldn't cope with the first bad thing. TV hurt – it was relentless, a misery factory, a vector for confusion and anger – so people sought escapism. Netflix boomed, but there's a limit to how far pixels can be a prescription for extreme anxiety, so as the things to do within the four walls of our houses rapidly exhausted themselves and the days became longer, warmer and increasingly full of a spring exploding into life, we went outside and, if we had one, into our gardens.

For the first time people noticed a male wren building one of its nests inside a shed, a gang of wasps girdling wood from the shed door or the characteristic sticky black poo of a hedgehog – yes a hedgehog – that had snuck through the compost heap and crossed the lawn in the dead of night. These encounters, however subtle or perhaps familiar to some, served as a poignant reminder that while our lives had been put on hold, the natural world was still springing and the Earth was still spinning.

UNDERGROUND FLIRTATION

Night crawlers wriggle beneath the soil's surface leaving behind little evidence of the fascinating lives they lead. When the leaves fall to the ground each autumn and slowly disappear, it's the subtle sign of their tireless work in the damp undergrowth. Globally there are over 3,000 species of terrestrial earthworm, ranging from the microscopic to the gigantic, the longest being the Giant Gippsland in Australia, which grows to a whopping 6.6 ft! Earthworms can be found anywhere that soil moisture and temperature allow, but they prefer clay soils that retain moisture and are high in nitrogen and phosphorus, so vegetable beds are an ideal hotspot to find these busy creatures. In the UK we have just 27 native species, the most abundant being anecic earthworms, which are large with a red-brown tinge. Each day we continue to find out surprising information about them – especially when it comes to their love lives!

Earthworms are hermaphrodites, meaning that they have both male and female sexual organs. Two individuals will share their genetic material during copulation, so ultimately each worm will end up being the paternal figure to some of its offspring and the maternal figure to others. If that's not mind-blowing enough, their pre-copulation flirtation game is something to behold. A 1997 study revealed that earthworms court each other by visiting one another's burrows up to seventeen times before mating occurs and, once it begins, it will last somewhere between 69 and 200 minutes. You could argue anthropologically that they go out on 'dates' before sealing the deal. Fred Sirieix – eat your heart out! Earthworms will have multiple sexual partners and research by scientists in 2013 looked at the paternal success of breeding pairs. They found that the first and third partners to mate with an individual would gain most, if not all, of the offspring's paternity. The second individual would have a very low success rate. The theory concludes that the first individual's sperm would naturally arrive first (as they say, the early worm gets the bird) and the third individual would displace the sperm of the second. So, if you're an earthworm expending a lot of energy visiting others' burrows, it's worth making sure you're not second!

With this novel or renewed interest in the world outside our windows it's no surprise that lockdown activities soon focused on improving our gardens for the direct benefit of local wildlife. As the daily cascade of social media posts displayed, gravel gave way to grass, and grass to wildflowers, bird feeders were filled and emptied and re-filled until the seed-sellers reported shortages, bug hotels and bee boxes were *de rigueur* DIY for many home-schooling parents, some stretching to boxes for their new hedgehog friends. The mission for many was to wake and make life better for life, and so in millions of tweets and posts we saw the nation's gardens transformed into wildlife wonderlands. But maybe more than any other there was one creative project that took off and soared ...

In 2017 the then-executive producer of *Springwatch* Chris Howard came up with an idea to make some two-minute 'Wildlife SOS' films for the series – very, very easy, cheap but definitely cheerful things anyone could do with next to no resources, which would definitely work to improve their patch. One was to sink a washing-up bowl into your lawn. Add some pebbles, a ramp in and out, some 'water weeds' and hey presto: pond. At the time no one complained, but it didn't seem to take off as a concept; maybe a bit too 1960s *Blue Peter* for 'pond snobs'. However, very fortunately, when we filmed it, we dug the plastic tub into Chris and his wife Laura's own garden and of course they nurtured it. Fast forward to the Self-Isolating Bird Club on its mission to engage with everyone loving the ordinary in the strange spring 2020. We asked Chris to give us an update on the proto-pond and all its visitors, and his contributions were brilliant, entertaining (with neighbourly interruptions and a swooping sparrowhawk) and informative, but also proved that there was proof in the pudding – the little pool was working. Newts, frogs and bathing birds had all followed Kevin Costner's immortal promise that 'if you build it, (they) will come' – not so much a field of dreams as a pool of promise.

People loved it. Within days our Facebook and Twitter streams were deluged by ecstatic 'mini-pond entrepreneurs', their washing up replaced by amphibians, dragonflies, damselflies, thirsty foxes, parched hedgehogs and, best of all, smiling children. And young people who built their own. And of course washing-up bowls are the 'gateway pond', after which things can rapidly get more serious. Even Chris and Laura's modest professional introduction had given way to a bigger pond, from which frogs and toadlets emerge en masse, badgers slurp and that evasive sparrowhawk swoops to bathe or stoops on those smaller birds who dare to try. Elsewhere boring corners of gardens were wetted; we saw raised ponds, uni-box ponds, dustbin ponds (a bit deep – be careful) and proper and posher ponds of all sizes and shapes. The gardens of the UK got wetter for wildlife. Superb, and effective!

Ponds instantly increase biodiversity as they can support two-thirds of all the UK's freshwater species. Some of the organisms you will offer a home to are common frogs and common toads; the newts, including the glorious great crested; the voracious diving beetles, which will predate the 'poles' of the aforementioned amphibians; damselflies and dragonflies, which will travel great distances and thus find even city centre water sources; and an abundance of other invertebrates that can appear seemingly out of nowhere. Many years ago I was helping to put the finishing touches to a garden pond, fixing the liner and shovelling some gravel onto the shallow sloping sides. When we stood back to admire our work it was a bit disappointing – new ponds look raw and plasticky – but we threw the hose into the shiny pit, turned on the tap and then went indoors for a cup of tea. When we returned less than an hour later it was half filled and jumping around on the surface was a pond skater. What a pioneer!

As a very young child crawling up to the edge of our pond (not a washing-up bowl, but my old baby bath), I was entranced by the astonishing diversity of form in all the organisms that lived in such a small body of water. Tiny *daphnia* or 'water fleas' flicking in clouds, like micro balloons; the vigorous super-fast sculling of the back swimmers and water boatmen; the fearsome patience of the alien water scorpion or the peculiar pumping of my favourites, the rat-tailed maggots whose long 'tail' siphons allow them to breathe through their bottoms before they emerge as hoverflies. For the young naturalist this was a whole world in miniature, a constantly evolving saga – tadpoles grew legs, pond plants flowered, blackbirds came to bathe and no two days were ever the same. I always wanted to get back to its shore for more, it was *Game of Thrones* years before its time, endless entertainment all year round, and only rain would drive me indoors to *Stingray* and *Thunderbirds*. Now, as an adult, I still stare beneath the surface of ponds and submerge myself in another dimension, and always emerge having observed and learned something new. In the past 100 years it's estimated that over one million garden ponds have been created and, although the number excavated throughout lockdown is currently unknown, judging by the images posted on social media it must have been in the tens of thousands. It's heartening to imagine that Chris and Laura, and the SIBC, played a small role in that, and that there might be a small child out there somewhere, prostrate on a lawn transfixed by a greasy grey maggot breathing out of its arse. Job done.

SEEING LIFE IN THE FAST LANE

When we experience the environment, we are confined to the limitations of our own senses and minds. We see, hear, smell and feel things differently to other species, so it can be really challenging, if not near impossible, to imagine the world beyond our own understanding. Sometimes though, science steps in with a new discovery to help us bridge that gap just a little. The very first eyes evolved over 500 million years ago during the Cambrian explosion and were simple eyes made of photoreceptor cells. Today they're a little more complex.

Humans have trichromatic vision, meaning that there are three receptors within our retina that help us to perceive colour due to their sensitivity to green, blue and red wavelengths, while most mammals have dichromatic vision, meaning they lack red-green cones and can see a blue-yellow spectrum of colours. It is believed that primates evolved trichromatic eyesight to better detect ripe fruits against the green forest backdrop. But, if you think our eyesight is good then you're in for a treat here!

Dragonflies were one of the first winged insects on Earth, evolving 300 million years ago, and evidence shows that their eyes have remained largely unchanged. Their eyes are compound, meaning that they're composed of different facets called ommatidia, which each have their own lens. Humans only have one lens in each eye, but dragonflies have 30,000 ommatidia, which means, yep, they have 30,000 lenses. This creates one large mosaic image, but that's not the best bit. While we only have three colour receptors (depicting green, red and blue), dragonflies have somewhere between eleven and 30. They're able to see in ultra-multicolour and even view the world in ultraviolet and polarised light at up to 300 images per second, whereas we see only 60 images per second. This super-fast, super-colourful vision gives them the ability to make fast judgements on the speed and direction in which to travel to successfully lock onto a single prey individual in among a swarm of flying insects. Who knows what an ultra-multicoloured world would look like? Maybe one day there will be the technology, but for now I find it really exciting to know these different dimensions exist beyond the realms of our imagination. There is something quite calming about that.

Freshwater ecosystems, including wetlands, lakes and rivers, might only cover 0.8 per cent of the Earth's surface but over 10 per cent of all life on Earth is dependent on them and they are home to 35 per cent of all extant invertebrates. Consequently these environments are one of our most diverse habitats, yet they remain an understudied and underappreciated environment. What is your garden bird list total? How about butterflies? Moths? Mammals? You might have some idea. And I bet you could identify the vast majority of UK mammals, but how many freshwater fish could you reliably put a name to? And how is your 'fish list' coming along? You see, as terrestrial, diurnal, air-breathing mammals ourselves, those beautiful, fascinating and important members of our ecological communities are out of sight and therefore out of our minds. Perhaps that's why fewer tears are shed over the fact that we have destroyed 90 per cent of our freshwater habitats since the industrial revolution. Now urbanisation, over-abstraction, and surges of pollution from intensive farming and the discharge of sewage are major causes of their ongoing declines. Over 10 per cent of the species that rely on these environments are now threatened with extinction. But while ponds may be declining in numbers they remain one of the most widespread habitats, occurring on every continent from the freezing Antarctic tundra to the steamy jungles of the tropics. And even further afield – ancient salt ponds have been discovered on Mars and share some very similar properties with those on Earth. It's strange but it could be true that ponds probably have intergalactic importance for life!

Whether ponds are repositories of extra-terrestrial organisms we can only suppose, but they have certainly been important resources for the evolution of life on this planet. Research over the last 20 years by the Freshwater Habitats Trust and the European Pond Conservation Network has revealed that ponds have significantly more biodiversity than rivers or streams and are home to many specialist species. One of my favourites lives a few miles from our home in the New Forest and they've been around a while. Tadpole shrimps, or *Triops*, are like miniature three-eyed horseshoe crabs, and are considered 'living fossils' as they have remained largely unchanged for 250 million years. In fact, the north American species *Triops longicaudatus* is one of the oldest animal species still in existence, first appearing in the fossil record more than 70 million years ago. This means that they were scooting about in temporary ponds when *T. rex* ruled the Earth. I love that! The UK species can only be found here and in the Solway Firth in Scotland and therefore they are rare and vulnerable, but I've kept their close relatives in tanks many times. You can buy their eggs online and, when they hatch, you can feed them on virtually anything – chunks of carrot are a favourite – but don't let them get hungry as they are ferociously cannibalistic and you'll end up with just one big fat 'dinosaur shrimp'. And please don't be tempted to try your own *Jurassic Park* by releasing them into your pond; remember these are non-native species ...

Of course as well as being great repositories of life, ponds are beautiful environments too, and not just aesthetically; remarkably, they are part of the key needed to protect the climate for future generations. You see, ponds are fantastic carbon sinks. In 1994 some small lowland ponds were built on Hauxley Nature Reserve, a former open cast coal mine in the north-east of England, with the purpose of attracting colonising plants and invertebrates. That worked, but after 20 years scientists noticed that a layer of dark, rich sediment full of organic matter had accumulated. This material differed noticeably from the clay soil below and using this sediment they were able to calculate the ponds' various carbon burial rates. Each square metre of pond could bury somewhere between 79 and 247 grams of organic carbon each year, averaging at approximately 142 grams. Admittedly that doesn't sound like much given that small ponds currently only occupy about 0.0006 per cent of the UK's land area, compared to woodlands at 13 per cent and grasslands at 36 per cent, but this carbon fixing rate is much higher than either of these habitats. And the humble pond's climate crisis mitigation doesn't end there.

The 'Urban Heat Island' phenomenon is well known and describes how the man-made concrete jungles of skyscrapers, roads, pavements and houses lock in heat as it bounces from surface to surface, raising the temperature of the air and all the objects and organisms within its boundaries. At the peak of summer London can be up to 10 degrees Celsius hotter than the rural areas that surround it, causing a loss of sleep quality and an increase in the chances of developing serious respiratory illnesses. As a result of these locally warmer temperatures, spring arrives earlier and summer hangs on longer, and sensors reveal that the plant growing season is on average elongated by a full five days. As Billy Idol proclaimed in 1982, it's 'Hot in the City', and even ponds located here are on average 3 degrees

Celsius warmer. But new research conducted in Belgium indicates the presence of these hot little ponds could be really beneficial as they can reduce the temperature of the surrounding area by two or three degrees, especially in the morning when evaporation is at its highest. Unfortunately, there tends to be a deficit of ponds in the hearts of towns and cities, which is a shame because these open-water bodies can quite literally help turn the temperature down.

Now, if that isn't a convincing catalogue of very good reasons to tempt you to excavate your Generation X albums, shift some soil and dig for a plethora of environmental, educational and aesthetic victories I don't know what is. And while you've got the spade out ...

The UK has been turning a blind eye to the catastrophic destruction of its wildflower meadows. Since the 1930s, we have foxtrotted and tango'ed, jitterbugged and mambo'ed, twisted, popped, pogo'ed and twerked as 97 per cent of this wonderful habitat has been torn from the countryside. Aside from all the plants, butterflies, moths, dragonflies, damselflies, bees, wasps and beetles that have all disappeared from this floral wonderland, many mammals and birds have too. Our few remaining native wildflower meadows are vital havens for biodiversity, and support those pollinators which service some 35 per cent of the world's human food crop. Wildflower meadows on agricultural land can also increase the crop yield of neighbouring fields by 10 per cent as soil quality is improved and the abundance of small mammals and birds helps reduce 'pests'. Indeed, farms with wildflower strips see a 40 per cent reduction in 'pest'-related crop damage. Given their rarity it's sadly unlikely that many people ever get to walk through these little postcard-sized patches of the past. To lie down and listen to that symphony of buzzing, to smell that warm, rich, musty spice of the variety of plant life, to marvel at the crazy palette of yellows, whites, blues, pinks and reds that pepper the green. I've been there, but I was so conscious of its value that the idea of even treading, let alone lying, on this tapestry of rarity was too much – I just stood and gawped in awe. But it doesn't have to be an impossible dream because if you have a garden you can fight back, take control, make your own, and when you make your bed, you can lie in it.

Lawns are bloody boring. There, I've said it. And it's true. Why do people go to such extraordinary lengths to attain the perfectly manicured boring green lawn? They mow, chop, water and spray chemicals on these boring patches of just grass. They hate plantains, daisies, dandelions and moss, and God forbid that a fox should piss on it, or a badger dig it, or a mole put a little mound of mud on these boring plots of shorn monocots. 'No, grass, we will not let you get to your reproductive stage, you will not flower or ever set seed, you will be constantly sheared so you will only have your boring leaves.' It's horticultural control-freakery gone mad and the result is ... boring. On the other hand, having a wildflower or wild grass patch in your garden can make a huge difference to the richness of life you'll encounter on a daily basis. Your garden will instantly be less boring. So, let's go! Select a sunny and open area, in truth the bigger the better to get a wider range of flowers going, but any little patch will do. Your meadow in the making doesn't have to be flat; banks with an incline can be just as effective. Using some muscle power or with a turf-cutter, remove the boring lawn to get rid of the grass and then the first four or five inches of topsoil to reduce the soil fertility, as most wildflowers do better on poorer soils. Once the ground is bare, and you've ruthlessly persecuted any of the boring remnants of lawn that have tried to return, go ahead and sow your chosen native flower mix. A mixture of annual and perennial species will ensure you'll get lots of flowering in the first year, which is satisfying. For a dense, full meadow, plant approximately 5 grams of seed per square metre having mixed them with 15 grams of sand. The best sand is that which is fine and low in moisture; dry silver sand works a treat and helps you see where you have sown. Spread the mixture evenly over the area that was previously your boring lawn and gently rake it over to increase the seeds' contact with the soil. The best time of year for this is between March and April, but local terms and conditions apply, especially late frosts. Alternatively, you can sow in September if the soil is light and won't get too waterlogged over winter.

It's always best to plant native species, as they're conditioned to survive in our environmental conditions and, more importantly, our native wildlife has co-evolved to coexist with them. Studies indicate that pollinators find native wildflowers four times more attractive than non-natives. What you can best grow will depend on your soil type, drainage, aspect, sunshine hours, etc, so do some research into the requirements of each wildflower species before you purchase your seeds. Many producers offer advice and a variety of mixes to best fit your prospective area. And when it germinates and grows you will have to decide which type of meadow you want to culture based upon those species and a mowing regime. Advice varies: you can make the first cut after six to eight weeks of growth and then continue to do so every couple of months during its first summer, but once established your meadow will only need mowing between one and three times a year. I'm no gardener but even I have been able to culture some very beautiful and productive alternatives to the bloody boring lawn.

SCIENCE IS IN THE DETAIL

It's early spring and the warm weather has brought a queen bee out of hibernation early. The wood anemones are starting to sprout and the birds are flexing their syrinxes – the bird's equivalent to a voice box – in preparation for the height of the dawn chorus. It's a beautiful scene, one that brings comfort. You may be enjoying breakfast on your balcony or reading a book in the afternoon sun when something catches your eye. You notice on closer inspection that your garden plants have some unusual holes in their leaves taking the shape of a half-moon crescent. What would have caused it and why is it there? It's something you may spend a few minutes questioning, looking for the culprit, but then you will go about your day thinking not too much of it. To be honest, you wouldn't be alone; it's something we have all missed for a very long time until a very interesting discovery was made in May 2020.

Researchers at ETC Zürich University were researching how bumblebees would respond to the odour of different plants inside greenhouses when they came across something unexpected – bite marks on the leaves in the shape of half-moons. Initially they suggested that the bees were feeding on the fluid inside the leaves, but as the bees did not stay for any significant amounts of time and they certainly weren't taking any leaf debris back to their colonies, this hypothesis was quickly ruled out. The big clue came when scientists realised that bumblebees originating from colonies with less food were damaging the plants more often. When plants are stressed due to pressure from disease or drought, we know that they often flower earlier to produce pollen and reproduce at a faster rate. As it turns out, bees are able to trigger the plants' stress responses, which stimulates them to flower early and release their pollen. This is useful for the bees, which emerge earlier than the plants on which they depend. As a secondary experiment, the scientists cut holes in the leaves themselves to see if anything else was going on. The flowers on plants that had holes cut by bees emerged on average 16 days earlier than the control plants with no holes. The plants with man-made holes did produce flowers earlier, but not as fast as those damaged by the bees. Now the question remains, are there chemicals in the saliva of plant-eating insects that could prompt early flowering? It's very early days in this investigation but it could open many doors!

In 1803 the English poet, painter, printmaker and luminary William Blake wrote a poem about innocence and its juxtaposition with evil and corruption, entitled 'Auguries of Innocence'.

Perhaps its most famous lines are 'A Robin Red breast in a cage / Puts all Heaven in a Rage' – not actually a metaphor for the human enslavement of nature, but for human slavery and the impossibility of freedom. For me, though, the opening stanza has always been formative in my approach to connecting with nature:

> To see a World in a Grain of Sand
> And a Heaven in a Wild Flower
> Hold Infinity in the palm of your hand
> And Eternity in an hour

There was not a day during the lockdown I wasn't mindful of this, and reassured that to access the purest exultation that the natural world has to offer I didn't need to be on the plains of Africa, in the jungles of South America or the frozen north or south of the planet. I just needed to sit in my garden for an hour with a handful of dirt and a dandelion.

Patchy islands

Our landscape has been chopped, sliced, covered and left patchy. Impassable barriers slapped bang in the middle of what was once pristine habitat rich with biodiversity. We are instead left with horizons littered with the harsh edges of roads, farmland and fences. Wildlife takes shelter in the small pockets of habitat – or islands – that do remain and then becomes isolated, unable to disperse safely when the space becomes limited or the resources run out. These small, divided populations lack the ability to move freely and as such are more susceptible to disease outbreaks, inbreeding and natural disasters, like fires, flooding or climatic changes, which all leave them vulnerable to localised extinctions. In the 1960s, Robert MacArthur and Edward Wilson developed the concept of 'island biogeography', which describes two related main principles: species richness will be higher in larger areas and the species richness will decrease with increasing isolation. So, essentially, the smaller the islands are and the further away they are from one another, the less wildlife will be sustained. The edges or boundaries around these fragments let in additional light and wind, and in some cases pollution from noise or chemicals, so the quality of the habitat is significantly reduced.

Much of the UK's landscape is fragmented and many species struggle because of it. Hazel dormice are a prime example. They're 'cute' little rodents with big beady eyes and round ears that live almost exclusively within the trees and dense shrubs. Dormice very rarely travel across open ground and, as a consequence, don't disperse and colonise new patches of habitat easily. Researchers have been monitoring the population, which has fallen by 50 per cent since 1995 and is at serious risk of inbreeding and the loss of genetic diversity. Globally, conservationists have been trying to increase the connectivity of these fragmented islands by implementing natural corridors, buffers and stepping stones. Hedgerows, initially used for marking ownership boundaries, have now become a vital part of the UK countryside, supporting the movement of species, like the dormice, stranded in isolated environments. The best hedgerows are thick and broad at the bottom, consisting of a variety of woody hedge species, like hawthorn, field maple, hazel or blackthorn. In the case of the dormice, researchers are even building aerial runways to grant them safe passage over open ground to nearby habitats. Species-specific links like this, or even things like badger tunnels, can help displaced wildlife cross roads and small open distances.

Nature reserves can also provide some refuge if they are large enough to sustain those species with larger ranges. They become a pocket of protection. But what if I told you that there is an existing network of connected and untapped areas that could be rewilded and that you – we – can manage in our very own hands? That we could all help patch up the UK's broken landscape by creating our very own links? As mentioned, 87 per cent of people in Britain have access to a private garden, shared garden, balcony or patio. This land combined is greater than that of all the nature reserves, coming in at approximately ten million acres. Considering the UK is made up of nearly 60 million acres, our gardens make up a huge 16.69 per cent of all that land. If you're one of that 87 per cent then there's something you can do – but you can't do it alone. Perhaps your garden is already wildlife-friendly and is visited by bees, butterflies, wildflowers and birds. You may well have insect hotels, bird boxes, feeders, a pond, a water dish, etc, etc. And if you're really lucky, you might get the odd glimpse of a hedgehog or a resident fox. This year my mum created what can only be described as an insect castle or manor house – it's nearly as tall as me with five floors and packed full of pots, logs, rocks and substrate that would be perfect for all sorts of creatures. A few weeks passed and I received a phone call wondering why it appeared no one, other than a few solitary bees, had checked in. You see, it is so important to create that vital habitat for wildlife but if your neighbour, their neighbour and their neighbour's neighbour has nothing but decking and an artificial lawn then your patch of oasis – perfect as it might be – may be stranded and out of reach.

The Hampshire and Isle of Wight Wildlife Trust in partnership with Southern Co-op have initiated a project called Wilder Portsmouth that aims to support the communities and streets that come together and take action to enhance their gardens and local spaces for wildlife. Outside of London, Portsmouth is the second most densely populated city in the UK and its green spaces are scarce. Gardens are especially important in these areas where urban developments and people have taken over. But here, communities and streets have come together, with some helpful tips coming from the Wildlife Trust, to create a network of stepping stones through their gardens to help species move around freely. These 'wild streets' have helped connect the community as well as benefit their environment. The first wild street in Portsmouth was Francis Avenue, where residents have built a green bin shelter for growing plants, and they've also collectively added bird boxes and insect hotels and cut hedgehog highways in their fences (just in case!). Neighbours share seeds, pots and plants in their nature exchange stations. Laura Mellor is one of the residents; she says, 'the idea is to make Francis Avenue into a green corridor. We don't have big gardens or much green space but if everyone utilises the space they have, by using it for plants and creating green surfaces, we can encourage birds, bees and other insects in the area. It doesn't seem like much but if everyone does it it can make a difference.' This street's effort certainly has made a big difference as the idea is now being adopted in streets across the city.

It's a simple concept and model that could be applied anywhere. Wouldn't it be fantastic to see wild streets popping up in every city or town? You can find lots more information by looking up #WilderPortsmouth or by finding the Hampshire and Isle of Wight Wildlife Trust online. You see, a garden fit for wildlife is a beautiful thing, but it becomes so much more when we, as a community, break down some of those barriers. We've got to let life move.

THERE'S POWER IN A UNION

I've been fortunate to promenade along some of the world's most famous esplanades. Brighton's sea front (with the distant echoes of Mods and Rockers and ... rock), Venice Beach, California (in honour of Jim Morrison), Fifth Avenue, New York (because Audrey Hepburn makes me swoon and I wanted to breakfast after looking in Tiffany's window), the Champs Elysees (terrible, a ghastly Christmas market at the behest of Megan), the Via Sacra, Rome (I took Megs, aged nine, to the Coliseum and re-enacted gladiatorial combat – I won!), the Avenue of the Sphinxes, Luxor (Megs, aged ten, she won, so we had to stop archaeologising and get ice creams). But for all the glamour and faded glory of these exotic pavements, sidewalks and causeways, no stroll down a pedestrian precinct has stayed with me so vividly as a back passage in Belfast.

Wildflower Alley. It's heaven in a former hell. On a sunny morning in the summer of 2018, while Megs and I were on our UK-wide bioblitz, I stumbled sleepily out of our van and into a paradise of colour, imagination, creativity, conservation and sublime community passion for life. I'll never forget it, obviously; I was shocked awake by the power of people to do enormous good for wildlife and for each other. I feel tearful even writing this. It was an alley that was riven with social ills, drug abuse, prostitution, dumping of trash ... the residents lived in fear and a mess of needles, rubbish and abuse. They cracked (pun intended), formed an 'alley gate group' and put up some gates to keep out the reprobates, initiated their own big clean and started to dream. To start with they broke their own ground, literally, finding whatever scraps of bare soil they had. Many had none, but undeterred they strung hanging baskets, requisitioned pots and planters, and brought earth to their barren part of our Earth. They scrounged some seeds, got some compost, and planted. Now it's a jungle – in places it droops green, flowers splash a spectrum of brilliance over the brown, murals glow and messages are writ large about the past, the present and the future, and there is a brick road that leads through all the wizardry. The fencing is blue, turquoise, lime, even shocking pink. Old guitars sprout ferns from their silenced bellies, paintings of cows and birds and bees peep through trellises bedecked with a tangle of natives and exotics. But the key is unified individuality. Please get that, it's so important: unified individuality – everyone is doing their own thing, in their own way, but it's all fused to make a bigger utopia. The whole project exemplifies the power of people; how a community can utterly transform the space they own to recover and enrich their lives. So you can keep your Californian skateboarders, your ancient Egyptian arcades, your posh Manhattan shops. If you want to stroll down an avenue of the world's best community conservation you now know where to go: Holyland, Belfast. And you will get a better cup of tea and slice of Guinness cake to boot.

SUPERWORM AND THE PLASTIC APOCALYPSE

Our homes and environments have an infestation that we simply can't be rid of. It's plastic and it's one of the major threats to all environments having spread to all corners of the planet, from Mount Everest to the bottom of the Mariana Trench. Behavioural change and mass clear-ups are needed, among other solutions, but there's an unexpected ally when it comes to plastic eradication: superworms. No, they don't wear capes, bench-press trains or fly around at the speed of light, but they do have the pretty impressive super-ability to chomp through one of the most stubborn substances on Earth; a man-made substance that remains on the surface of our planet for somewhere between 500 years and one million years without budging.

In 2018, there was an estimated 359 million tonnes of plastic produced globally, including 33 million tonnes of polystyrene. Much is discarded after its single use, getting dumped in landfill, littered across our landscape or degraded into micro-structures. It's a plastic apocalypse. With little idea about how to remove such damaging waste from our environment, scientists turned to some unlikely super sources.

Scavenger worms have been around for over 100 million years, with prehistoric species eating everything from plesiosaurus bones to the maggots and mealworms we see today feasting on the organic matter in our gardens. Scientists recently tested the larvae of a superworm species called *Zophobas atratus* to determine whether it could successfully consume styrofoam, a form of polystyrene. To everyone's amazement, these worms had the ability to survive solely on this plastic due to specialised gut microbiota that depolymerised the long-chain molecules within polystyrene into carbon dioxide and products with a lower molecular weight. The idea for this experiment came about after scientists found that waxworms and mealworms could also ingest plastic, but as the larvae of *Z. atratus* was considerably larger (three to six centimetres long) they found that this superworm could eat four times more at 0.58 milligrams per day. If scientists are able to culture this gut microbiota independently from the worm larvae, they might be able to find a way to speed up the process of plastic biodegradation, and give us a fighting chance of turning the tide on the plastic apocalypse.

Community Conservation is a relatively new approach. The traditional method is to acquire land in as large a parcel as possible, put a fence around it, keep people out (or throw them out), and make it sacrosanct for wildlife and nature. This was the sort of practice that was pioneered when the first parts of the United States were protected – Yosemite National Park in 1864 and Yellowstone National Park in 1872. Both the early settlers and the indigenous people inhabiting these areas were removed from their land and displaced onto marginal areas surrounding the new wildlife reserves. The environmental thinker and activist John Muir, known as 'The Father of the National Parks' and 'one of the patron saints of twentieth-century American environmental activity', argued that 'wilderness' should be cleared of all inhabitants and set aside solely to satisfy our need for recreation and spiritual renewal. It was a philosophy that became a US national policy with the passage of the 'Wilderness Act' in 1964, which defined wilderness as a place 'where man himself is a visitor who does not remain.' This idealised preference for 'virgin' wilderness was thus enshrined in a conservation movement that bizarrely acknowledged no connection between nature and human culture. I admire what John Muir achieved, but struggle to find any peace with his ethics; he loved nature, but not human nature, and seemed blind to any wildness in humans. While he could get very romantic about Yosemite's landscape and ecology, he described the native residents as 'most ugly, and some of them altogether hideous'. And of course 'human wildness' was something that was rapidly disappearing – remember that these parks were established before the Battle of the Little Bighorn and Wounded Knee. Muir and other conservation trailblazers were saving wildlife while, just over the Rocky Mountains, their countrymen were all but completing a genocide of the continent's first nation peoples. In short, it shouldn't be surprising, but it's certainly sad that conservation has very colonial roots and, historically, an ethos that venerated the nobility of nature, but not that of the people who had helped shape it for thousands of years.

In the name of conservation it is estimated that up till now a bare minimum of 20 million people have been displaced from their lands around the world, 14 million in Africa alone, and India admits to evicting 1.6 million of these so-called 'conservation refugees'. Such forced expulsions continue even now. The Indian government turfed out 100,000 people in Assam in 2002, and it's likely that two or three million more will be displaced by 2030. And this is where things get decidedly dark and palpably disappointing. You see these wholesale evictions are a heavy-handed response to a 1993 lawsuit brought by the World Wide Fund for Nature, which demanded that the Indian government significantly increase its protected areas for tigers. When it comes to massive environmental and cultural destruction the usual suspects are the petrochemical giants like Shell, Exxon, Chevron-Texaco and BP, maybe the monstrous mining corporations such as Rio Tinto, and global resource harvesters such as Boise Cascade, Mitsubishi, Conoco-Phillips, International Paper and Weyerhauser. But what is truly shocking is that large conservation agencies are in the frame too ... We may be paying our annual subscriptions to one or more of these agencies, or if we have a bit of spare cash donating it to their multitude of appeals, but do you really think that our modest remunerations are paying for all their staff and global projects? Obviously not; necessarily the financial support for large-scale global conservation has moved into another league and very large and very wealthy foundations, a plethora of banks, governments and many of the huge transnational corporations are dropping cash into their hungry purses.

In the 1990s the United States Agency for International Development (USAID) – described on their website as 'the world's premier international development agency and a catalytic actor driving development results, (whose) work advances US national security and economic prosperity, demonstrates American generosity, and promotes a path to recipient self-reliance and resilience' – stumped up almost $300 million for international conservation. The five largest conservation organisations, including Conservation International and the World Wide Fund for Nature, received more than 70 per cent of it. But guess what? The world's indigenous peoples received nothing, despite the fact that, although they comprise only 5 per cent of the world's population, these people protect over 80 per cent of its biodiversity.

So these charities are big and wealthy and well connected, they have branches all over the world, millions of individuals like you and I are still dropping our pounds in their pots, and they have been busy. Yes, these big international conservation bodies have successfully led a charge to increase the number of protected areas to safeguard global biodiversity. In the early sixties there were about a thousand national parks and reserves; now there are ten times that number, covering 12 per cent of all the planet's land – an area greater than the landmass of Africa. This sounds like a laudable and remarkable achievement but then ... it hasn't worked, has it? Since the 1970s we've destroyed more than half the world's wildlife and, in Africa alone, where more people than anywhere else have been displaced to make the parks and reserves, a staggering 90 per cent of the remaining biodiversity lies *outside* these protected areas.

The unrelenting lust for energy, tropical timber and precious metals remains the primary threat to indigenous peoples all around the planet, but close behind is this old model of classical (colonial) conservation. And it may shock you further to learn that these two threats are not entirely separate, as some of the planet's largest conservation agencies are also working very closely with some of the most aggressive multi-national planet-destroying companies, many of whom form part of a Conservation International body called the 'Center for Environmental Leadership in Business'. Big conservation is seemingly in bed with big bad business. Their argument would no doubt be that if they were to get out of that bed they would lose access to millions in funding and access to be able to lobby these global conglomerates. What do you think?

Partly as a reaction to this questionable method of protecting wildlife and ecosystems, community-based conservation projects began to appear in greater numbers in the 1980s, initially following successful protests against the aforementioned roughshod attempts by international agencies to protect our dwindling biodiversity. The basic premise of community conservation is to integrate the improvement of the lives of local people with the safekeeping of wildlife, sometimes by creating protected areas like reserves. But while this inclusive, democratic and far more ethical strategy has produced many notable successes, large and small, internationally and within the UK, it has also seen a fair bit of controversy. Community conservation can be ineffective due to a lack of adequate resources, their uneven implementation and over-ambitious or inappropriate planning. Smaller projects in particular can suffer because they are often motivated and energised by fewer individuals, making them vulnerable to any changes in 'staff', and the subsequent loss of both passion and skills. And of course without the right skill set and necessary knowledge in the first place, such projects can be counterproductive in terms of real wildlife conservation. How many times have I been led through what had until recently been productive habitats, but which now had been 'tidied up to improve access'? Lastly, the tricky hypocritical bit: as much as decentralised democracy is a great idea, at a very local level it can be too easily derailed by the wrong stakeholders – people whose involvement is either ignorant ('let's tidy it up') or counterproductive through their vested interests ('let's not do anything').

What would pay dividends in so many cases is a partnership between the community, with all its local knowledge and traditional commitment to the place, and some up-to-date scientific knowledge and practical experience. In my experience this is far more productive in terms of increasing both biodiversity and the long-term care of any protected area. But it's not easy, particularly as centralised power resents devolving to local power and therefore often refuses any economic input. Which in turn means that many community conservation projects are self-funded and need an economic angle, and those economic angles invariably mean some form of development or resource exploitation, which equally invariably can countermand or compromise the conservation goals. Cue economy versus ecology, and conflict between the perceived priorities to produce those socio-economic benefits, and then the end of the harmonious unit, and often the project itself.

You may think I'm being cynical, but I'm not; I'm reporting what I've seen and I'm being pragmatic. Only a fool would think that there could be a one-size-fits-all local, national and global model for effective conservation; we've tried it with all those parks and nature reserves and that's failed. Even in the UK, where thankfully very few people have ever been displaced in the name of conservation, it's failed. As we will see later, our national parks are in some regards a national disgrace, and the vast majority of our state-owned and managed National Nature Reserves are in a parlous state through lack of investment of public funds. No wonder people and communities feel compelled to give it a go. My point is that we need to properly identify how and why such projects can succeed and use this knowledge to ensure future successes. Some research has been done and it reveals that such schemes must be robust enough to overcome challenges from national contexts, such as corruption and vested interests, and can be handicapped by ineffective local governance and unsupportive cultural traditions. But it also strongly suggests that when plans are well-designed in favourable local contexts, community conservation can be a very effective method of protecting people and wildlife. I think we may sometimes spend too much time measuring species success and not enough time analysing how we actually achieve that in human terms. Just saying ...

As the founder of the Cliddesden Community Conservation Group, Alison Mosson was very proud on the day that the 2018 bioblitz rolled up to the meadow the group own and manage. The former pasture field lies alongside the M3 at the edge of this Hampshire village and was purchased at auction in 2012. The group is very nature-focused and having formed in 2004 began to maximise the community's green spaces for wildlife, planting thousands of trees, conducting their own surveys, restoring the pond, putting up nest boxes and generally building a better and wider ecological awareness among all the residents. But I get the impression that the team were all itching to get stuck into a bigger project. As outlined above they had to do it all themselves – the seed was gathered locally and scattered collectively, and now 150 species of wildflower prosper in this rich, thick rug of colour and life. Tea was served, the butterflies bounced, the bees buzzed and the team of assembled experts were busy, finally clocking up 386 species for the day – very impressive. I didn't see anything too exotic, but I did see a conspicuous community accomplishment that had clearly made a positive difference for both the people and their wildlife. Top work!

WATCH OUT FOR THAT BRAMBLE

We are entirely obsessed with making our wild spaces look 'neat' and 'tidy' – whatever that really means – and one of the most common plants to be ripped up or torn back is bramble. It is an invasive plant that will outcompete wildflowers on verges, and therefore does need to be managed in some areas. But it's also a plant we have a strong history with. Bramble always reminds me of my grandmother who used to take me to the alleyway near her house with big empty ice cream tubs to collect the blackberries growing on an unruly and forgotten patch of bramble in urban Southampton. We'd fill our pots, take them home and I'd help as she prepared the most delicious crumble on Earth. Like all young children, I'd never be able to resist pinching a few berries as we went along. Some were so sweet and juicy and others so sour they'd leave your mouth tangy ... it was like a game of blackberry roulette.

There are over 400 bramble microspecies in the UK and this accounts for the vast variation in berry taste. Archaeologists have found evidence of their seeds in the stomach region of an 8,000-year-old Neolithic man excavated in Essex, so humans have been snacking on its fruit for at least that long.

There is some interesting folklore surrounding the bramble plants. Legend has it that Michael, the most holy of angels, fought, defeated and banished Lucifer from heaven. Lucifer fell from heaven, transformed into the devil and then landed on top of a bramble bush on his entry into hell. He was so mad that he spat on the bush, cursing its fruit from 29 September, which is apparently the date he fell. So in medieval times, people used to plant brambles on graves to keep the devil from getting out and to keep the dead in.

All our cultural associations with brambles seem very dark, but in reality these plants support so much life. You can find bramble in hedgeland, woodland, scrublands and wasteland, and it provides refuge for hedgehogs, dormice and the birds who build their nests within its woody branches and feed on its fruit. In the right area, brambles can be such a great addition to the landscape, promoting biodiversity and encouraging our grandmothers to make some of the best crumble and pies around!

Sometimes people don't need to own land to protect it. Take road verges; they are inevitably managed but not always owned by the Highways Agency, a sort of quasi-mysterious body that seems inaccessible and autonomous and, when it comes to mowing verges, devils incarnate to some. I'm not sure that's the case, but what I am sure about is that with the loss of most of our traditional meadowlands through post-war agricultural intensification, the UK's largest nature reserve is now potentially represented by roadside verges. Yes, those ribbons of green, which lace the length and breadth of Britain, that we spend more time driving past than looking at, extend to perhaps half a million acres, which is 780 square miles, which is a massive area and if managed correctly a very important resource for wildlife, particularly wildflowers.

Flowers are cut a rough deal. We swoon over dormice, pine martens and badgers; we rush around feeding and counting birds; we survey and worship our butterflies; dragonflies and damselflies are hip these days; and, yes, even bats get more attention than the very things that, ultimately, sustain them all. And I was a culprit of ignoring botany too – until my mid-twenties I was a 'daisy and dandelion' kind of ecologist. Then I embarked on a UK-wide ramble, ostensibly to see all the UK's orchid species (yes, I know, still preoccupied with the exotic) with a brilliant, then-young, naturalist called Andrew Welch. He had already sharpened his green nose while mine had been in kestrels' nests and down badger setts. I can't say he taught me all I know, because I've learned more since, but his enthusiasm for plants was enlightening and essential to better round out my appreciation of the UK's terrestrial ecologies.

(As a quick aside, we furtively researched the loss of the summer lady's-tresses, an orchid last recorded in the New Forest close to where we lived in 1952. We were both born in 1961 and it seemed so tangible still, so for several summers we scoured the last recognised sites, chasing away the burgeoning numbers of ponies to peer into all those damp places their ravenous grazing could not reach, under fallen trees in the fringes of brambles, but unfortunately the post-war drainage had been too effective and while the ponies prosper the orchid has gone. Anyway, back to the verge.)

During lockdown the mowers and flails were grounded. Hallelujah, hallelujah, hallelujah! It was like a dream come true; the grasses grew, the flowers too, and bees and butterflies flew. And how we loved it! Yes, at last everyone realised that each spring those otherwise shorn and browned-off strips could be a riot of colour, of joy, of life. Social media was ablaze with people at first rejoicing and then pleading with councils and highways just not to make the cut. And then as the lockdown eased, the screaming started as the contractors' contracts needed to be satisfied and the swathes of emergent richness were unceremoniously slashed and torn. But not all of them! No, the message reached out through the ether and some councils responded so brilliantly, adjusting their mowing regimes and actually planting those spaces with flowers. Superb, progress, a move to not mow in the right direction. Okay, some of the roundabouts and central reservations were not planted solely with native species, and some were planted with flash-in-the-pan annuals, but at least people out there were trying, giving us a starting point to engage and say, 'thank you, not bad, how about trying this next time'. Unfortunately, the ultra-green far righteous botanists failed to see such an opportunity to encourage improvement and all kicked off very publicly about non-native cultivars, and accurate but unwarranted pedantry around management. A shame; they know their plants and their sentiments were grounded, but they know little about diplomacy and how to instigate progress and they lost at least one friend in the process.

'Say No to the Mow' is an initiative run by the excellent Plantlife charity. It encourages us to choose a patch of lawn and simply let the grass grow, just spare it from your mower's blades until the end of August. Less work, more flowers, a lot more insects ... in fact Plantlife's botanical specialist Trevor Dines says that a mere eight dandelion flowers provide enough nectar for 15,000 bee visits per day. The 'No Mow Zone' should be as big as you have the space for, but even a single square metre will make a difference, particularly if you can convince your neighbours to do the same. This year, given everyone's outspoken enthusiasm for a bit of rough on the roadside, Plantlife modified the idea and launched 'No Mow May', so just no cutting until June, but you could try this in any of the spring months, or even all year round. Trevor recommends that the best haircut for your lawn is a Mohican, whereby some grass is left to grow tall and spiky, but the rest is cut shorter and gets trimmed once a month. This advice followed their 2019 'Every Flower Counts' citizen science survey, where people were asked to count the flowers in their lawns. The results were a little counterintuitive – you might think that that lawns left to grow tall and raggedy would offer the best resources for pollinators, but in fact it's actually shorter lawns with flowers like dandelions, daisies and clover that produce the most nectar and the greatest density of flowers. This is because these species are small and squat; they creep across the soil and can survive being mown, something that can stimulate them to produce even more flowers.

But variety is the spice of lawn life, and thus all short is not the best option. In the spiky part, your Mohican, the types of plants that prosper flower for longer and more species can establish themselves. These stands also offer shelter , while the short-back-and-sides produce all those extra flowers. The punky part should be cut just once in late August or early September. This allows the species there to produce and drop their seeds ready for next year. My four Mohican bits get strimmed in September, then I leave the cuttings on site to ensure all the seeds fall before raking them up a couple of weeks later.

Combined, the flowers growing in the lawns surveyed produce a whopping 23 kilograms of nectar each day, enough to support 2.1 million honeybees – the equivalent 60,000 hives. Better still, 80 per cent of lawns supported around 400 bees a day on the nectar produced by dandelions, white clover and selfheal. Even better still, the remaining 20 per cent of 'super lawns' were found to be supporting ten times as many; that's a massive 4,000 bees, not to mention the enormous diversity and abundance of other pollinators.

Now, leaving patches of a boring old plain grass lawn to flower and grow rank is one thing, but once hooked on your mini-meadow you can go further. Lawn grasses are ferociously competitive plants, so it's very hard for other species to establish under their ruthless regime. If you scatter seed it may germinate, but the little seedlings just won't get through the aggressive sward of grasses. You can see where I'm going here: the grass has got to go, roots and all. The first time I went all-out wildflower meadow crazy I de-turfed the lawn by hand, with fork, spade and shovel ... on subsequent ventures, I've hired machinery. If you get back to bare soil then your little seedlings will grow and your plant biodiversity will instantly soar. I put in a mix of annuals, to get it going in year one, and perennials to take over after that, and if I'm out and about and scavenge some seeds I scrape a little patch of clear soil and give them a go. I've found that a mini-meadow like this this will last between five and seven years before the grass gets back in and chokes it up. If you can get yellow-rattle, a grass parasite, to establish this can buy you more time before you need to go back to bare-earth basics again. Good quality, locally sourced and organic wildflower seeds with high germination rates are not cheap, but from my experience you get what you pay for. And this year my Mohicans paid dividends; filled with a colourful palette and ripe with nectar they became a magnet for hundreds of species, and with all the dock, teasels and thistles we've still got those little flocks of goldfinches to look forward to.

Megs and I did a live Big Butterfly Count on our Self-Isolating Bird Club streams and focused on one of the Mohicans. It was a chilly morning, not the peak time, but we still scored six species in our fifteen minutes. When the sun shone it was alive with whites, browns, skippers and even some blues, a miniature festival of fluttering insects

– absolutely joyous! As has been the response to Butterfly Conservation's simple but ever-impressive nationwide citizen science survey.

The Big Butterfly Count began in 2010 and rapidly became the world's biggest citizen butterfly science project. It runs from the middle of July until early August, the time when most butterflies are in the adult part of their lifecycle and thus most easily seen. You just have to find a (preferably sunny) spot to spend fifteen minutes. Anywhere will do – gardens, school grounds, footpaths, nature reserves, city centre car parks – as long as there's something like a buddleia bush to attract the insects. Then count all the butterflies that you see that correspond with those on the downloadable chart or the accompanying app – fifteen common and easily identifiable species plus a couple of day-flying moths. You can do it as many times as you like; just submit separate records for different dates either at the same place or for different places, and remember that your data is useful even if you don't see any butterflies or moths. Knowing where things aren't is as important as knowing where they are. This year a staggering 142,976 counts were submitted by 110,733 participants and they counted – get this – 1,508,794 butterflies! As I write it's still early days in terms of data analysis, but large whites narrowly pipped small whites into top place, with gatekeepers, peacocks and meadow browns following in that order. And the organisers have reported that they saw a 200 per cent rise in visits to their website and thousands more messages than usual on their social media channels from people looking to find out more about the butterflies and moths they were noticing during lockdown. Superb!

GIVE IT SOME HEART

We all have butterflies in common. Whatever your age, your career, your interests in natural history, the emergence of the first butterfly of the year is something that we all look forward to. It's their beauty and their grace that leaves us in awe. As children I am sure we have all followed them around, going from flower to flower on our very own garden expeditions. It leads to a lifetime of admiration for these colourful beauties. Their wings are essentially made up of two transparent membranes covered in scales fused together that act as light vectors, absorbing and reflecting different wavelengths to give the illusion of various colours. This is a phenomenon known as structural colour, and is widespread throughout birds and insect orders.

Contrary to prior beliefs that butterfly wings are just lifeless membranes, scientists looking at hairstreak butterflies in January 2020 discovered that underneath their wing scales is a network of living cells. Looking a little closer, they noticed the movement of haemolymph (the invertebrates equivalent to blood) being pumped through veins in the wing. It was believed that haemolymph was just used to inflate the wings once the adults emerge from their chrysalis, but we now know it is present throughout the butterfly's lifespan. It's pumped by a 'wing heart' that propels the haemolymph around the wing several dozen times a minute. The reason behind this spectacular adaptation is to regulate temperature. Given how thin their wings are, butterflies could rapidly overheat in the hot summer sun, or get too cold in cooler environments, which would end or impair their flying abilities. The thermal images from the study are beautiful, showing the patterns of the veins within the wings and the red heat they give off. On average the living areas of the wing were 10 to 15 degrees Celsius cooler than the non-living parts of the membranes. There is something so beautiful about finding new adaptations in the animals we encounter often, and the concept that butterflies have hearts in their wings just makes them that much more magical, don't you think?

At the end of January the RSPB runs its Big Garden Birdwatch (BGBW) over a weekend. This survey started way back in 1979 in partnership with *Blue Peter*, when it successfully solicited 34 mail bags of results. In 2001 it welcomed parents and became a family event, and now over 40 years later it is the world's largest wildlife survey with around half a million people studiously watching their bird feeders for an hour of recording, some drinking vegan hot chocolate and sharing biscuits with poodles ... I like it because you can do it in your own kitchen or lounge 'hide', you are counting 'your' familiar species and there is always a slightly competitive edge. In my case a highly competitive edge.

For the last few winters I have been fortunate to have had a pair of willow tits visiting my feeder at this time of the year. Now, as they've declined by 94 per cent since 1970 due to habitat destruction, having these little beauties on my patch is a real coup. Did they show up during that hour spent surveying? Did they hell! Birds eh, can't live with 'em, can't live with 'em. Maybe next year.

Nearly nine million hours have been spent by participants of the BGBW counting their garden birds, and the total number that have been recorded is in the region of 137 million. Needless to say, this has generated an awesome data set and revealed real trends in changes in the abundance and distribution of garden birds over the last 40 years. Indeed, it was this survey that alerted us to the decline in song thrush numbers as the species was in the top ten at the BGBW's outset. By 2019 they had declined by 76 per cent and were only just propping up the top 20. House sparrows have also dramatically decreased, dropping by an alarming 56 per cent, and starlings by a shocking 80 per cent. Winners include great tits – up by 68 per cent – and long-tailed tits – up by 44 per cent. Blackcaps have become more regular too, as some are now spending the warmer and wetter winters here in the UK rather than migrating south to Iberia and North Africa. Exciting vagrants spotted by participants have included an American robin, a black-throated thrush, a common rosefinch and a yellow-rumped warbler, which showed up in a very lucky garden in Durham.

We can't all have yellow-rumped warblers, or bloody willow tits, but we could all have more house sparrows and starlings if our 'built environment' was a bit more wildlife friendly. Of course, many of us reshape whatever we've got garden-wise to improve its potential, but it can be a real uphill struggle for those living without outside spaces, especially in newly built properties. Our homes could and should be an environment we can sympathetically share with wildlife; they can offer so many resources.

Years ago, because I couldn't afford a house in the UK, I bought a semi-restored, still semi-crumbling farmhouse in

France. When I moved in it was already inhabited ... by bats, dormice, martens, barn owls, loads of lizards and a very cheeky pair of wrens who were nesting in the lounge. I'm certain most people would have blocked up all the holes and swept up all the mess, but I didn't – and never regretted it. Okay, when the martens chased the dormice around the loft the poodles' early warning systems went into very loud early-morning overdrive, but lying watching the Champions League final with a bat detector going berserk on my chest as all my Serotine bats exited to avoid the penalties was sublime. And yes, naturally there was poo, but there was a dustpan and brush and disinfectant too. Of course this was a very old house – it had holes – but new builds are designed to be hermetically sealed; there is no way in for birds and bats and bugs. But we can retrofit bird and bat boxes and bug hotels, if we are interested in wildlife, but sadly we are still a minority, and those who are not interested don't bother, and again sadly they are still the majority. Thus species that have adapted or evolved to rely upon our homes for theirs have lost out. And swifts are the classic example.

Swifts are totally amazing birds that travel over 14,000 miles a year and return to the UK to nest, and rest. For the remainder of the year they eat, drink, sleep and even mate in the air, but when they return here they are meant to be able to swoop up under the eaves and find a dark spot in a roof space to lay their eggs. But now they return, swoop, and smack straight into the gap-free plastic soffits that clad every new property, leaving them helpless and homeless. A recent survey of 3,400 swift nests found that nearly 80 per cent were in houses built before 1919 and that since the mid-nineties swifts have declined on average by more than 50 per cent across the UK.

I have to be honest, the loss of nest sites is only one factor driving this enigmatic species' decline; the catastrophic destruction of flying insects through pesticide abuse and issues across their huge migration pathway are likely the primary factors involved, but lack of nest sites is contributing and we can do something about that. You can buy swift nesting boxes and builders can buy swift bricks, which they can integrate into pretty much any new project, such is their variety. Do they work? Yes, definitely. Are they expensive? Looking online at one-off sales to individual consumers they range from £4.95 to £165 per unit. A good tried-and-tested one is between £25 and £40, so not dirt-cheap, but research in 2020 shows that the average new-build house across the UK costs £293,167 and consists of on average 5,180 bricks, meaning the cost per brick on a new property is £57. Of course, individual bricks come very much cheaper, but in terms of what we pay, that's it, and thus one £25 brick seems quite reasonable. So why aren't they being used in every new build going up? Well, guess what? It's because they don't have to be.

But some people are trying. Manthorpe Building Products have been making their pioneering swift bricks in conjunction with the RSPB and Barratt Developments Plc. It has been well researched and is well constructed and their use has been slowly increasing in recent years. Among the first of Barratt's building projects to pioneer these bricks were those at Ottery St Mary near Exeter and Kingsbrook in Aylesbury Vale, where of 2,450 new homes nearly 200 had swift bricks fitted between them. These developments were built with nature in mind as they also included orchards, hedgehog highways, newt ponds, and bat and owl boxes. It's a start, but why – given the cult status of the bird, its terrifying decline, the fact that we have one of the answers sitting cheaply on the brick-shelf right now – isn't it mandatory for builders to use these in all new-build projects, as well as bat bricks, bee bricks, starling and house sparrow bricks?

If we want our communities to be better places for us to live, then we need to ensure we share them with nature. We need a better grasp on how community conservation can be effectively performed, we need to actively encourage the people who manage and shape those physical spaces to understand how to do so to maximise their biodiversity and to take advantage of new solutions when they become available. As is already becoming clear, we – that's me and you – need to take control.

How green are your actions?

In the last ten years it has become increasingly cool and even fashionable to live an eco-friendly lifestyle. A cross-generation study conducted in America and Australia found that 93 per cent of people were concerned about the environment – namely biodiversity loss, climate change and plastic pollution – and a huge 77 per cent of individuals wanted to learn how to live more sustainably. I like those statistics a lot. We are striving more and more to think critically and reduce our impact on this planet. It is no longer socially acceptable to chuck your rubbish on the ground, rely solely on single-use plastic water bottles or purchase a pair of jeans that will only ever be worn twice. I don't know about you, but when I forget my reusable water bottle I feel tremendously guilty if I end up having to purchase a single-use bottle. If I do, I make sure to repurpose it in some way to ensure I get the most out of it, but even so I feel bad. If I forget my 'bag for life' at the supermarket, I end up with a boot full of loose vegetables rolling around because I cannot bring myself to buy any more bags that I know I have in abundance at home.

I have always been conscious of these things, as you'd expect growing up with Chris in my life, but there has definitely been a shift in the previous few years. Whenever my teacher talked about taking action to protect the environment at school, it was always in reference to the three Rs: reduce, reuse and recycle. Recycling is pretty commonplace these days, but many have also expanded on these actions to better the world around them by taking things into their own hands. Beach cleans, sustainable fashion, cycling, volunteering, signing petitions or even engaging with your local MP or MSP, etc. But how effective are these actions? Other than making us feel good about doing something positive, how much of an impact can they have?

Humans are naturally drawn to the water. Studies show that being near a water body – the ocean, rivers or lakes – has a positive impact on our minds, boosting creativity and lowering anxiety and stress. This phenomenon is referred to as the 'Blue Mind' and as temperatures hit highs of 32.6 degrees Celsius this summer (2020), thousands of people seeking to cool off and relax took to the beaches. Other than the serious concerns about the spread of Covid-19 due to the lack of social distancing measures, beaches were left in a state beyond belief. Dorset council said they collected 33 tonnes of discarded rubbish between Bournemouth and Poole beaches over the course of a couple of days. If not collected, all of those tonnes of plastic and other waste materials would have ended up in the ocean, and much probably did despite the council's efforts. Already each year an estimated eight million metric tonnes of plastic enters the seas. The impact of this pollution has been widely reported as microplastics enter the food chain and cause toxic damage. In December 2019, a sperm whale washed ashore with what was described as a 'litter-ball' of plastic waste in its stomach weighing 100 kilograms. Off the coast of Australia, seabirds have been found with 8 per cent of their total body weight made up of plastic. And as micro-plastics bioaccumulate in fish and other organisms, the average human now ingests up to five grams of the stuff every week – the equivalent of a credit card.

We are our own worst enemy. We are drawn to the water for its beauty and its calming properties, and yet we trash it to spite ourselves. But fed up and kicking back, communities and individuals are picking up bin liners and buckets to go out and clean the beaches themselves. This not only improves the coastal and marine ecosystems, but also provides a valuable opportunity to gather data about the state of our coastlines. The Marine Conservation Society (MCS) initiated the Great British Beach Clean, which has been running since 1993, and to date they have collected 319 tonnes of rubbish. All data is gathered and contributes to a worldwide report on litter levels, which is used to help inform new policies, such as the ban on microplastics in cosmetics, better labelling on wet wipes, the ban on plastic straws and the plastic bag charge. Of course, you don't have to wait for this event each year to get involved. Imagine if every person was to pick up at least one piece of litter each day – that would mean 7.8 billion pieces of rubbish would be removed from our landscape. What a better place the world would be. I hope one day we will learn to take everything and leave only footprints and sandcastles in the sand.

We also have a responsibility as consumers and many of us are wise to the impact our purchases can have when it comes to influencing the market and environment. Seventy-three per cent of eco-conscious buyers research the cleaning products they purchase, 73 per cent also look into cosmetics and 61 per cent research where their food has come from. Forty-seven per cent of internet shoppers worldwide have been found to ditch companies entirely if their products or services violate their personal beliefs. Clearly there is mounting pressure to be sustainable, so to meet the demands of our 'greener' shopping lists, manufacturers and brands are making some big changes. Fast fashion is the speedy production of cheap clothing that is sold by big retailers trying to keep up with the latest trends. The industry contributes 10 per cent of global carbon emissions and is the second largest consumer of water, with over 85 per cent of textiles each year ending up in landfill. For one single pair of jeans, 4,000 litres of water is needed, which is the equivalent of one person's drinking water for ten years. Ten years! But we've had a sudden awakening to this thirsty issue and retailers are starting to respond to our shift in demand. Here's a few examples:

1. *Levi recently committed to a collection of jeans made from rain-red hemp that reduces water usage by 30 per cent,*
2. *H&M has adopted ECONYL fabrics made from 100 per cent regenerated fibre from discarded fishermen's nets and pre-owned nylon,*
3. *Mango aims to purchase 50 per cent of its cotton from sustainable sources by 2022.*

There is a long way to go, but we must encourage and celebrate the brands making an effort to get there. Alongside ethical fashion, consumers are moving towards a host of eco-products, like stainless steel razors, menstrual cups, metal straws, wooden toothbrushes, cruelty-free makeup or reusable wipes and nappies. Your pounds have the power, so spend them wisely.

One of the other actions that Chris and I often ask people to participate in is to sign petitions and contact their government representatives to protest policies or environmental injustices. Whenever the situation crops up, I always notice comments on social media saying, 'I've done it but I am not sure it will ever get anywhere.' As it stands, a petition must have 10,000 signatures to warrant a reply from the government and 100,000 signatures for the topic to be debated in parliament. Cristina Leston-Bandeira is a professor of politics at Leeds University and has studied the effectiveness of petitions. She says, 'in general terms I can say that petitions are the method that has enhanced engagement with parliament over the last few years.' Regardless of political impacts, petitions and letters to MPs or MSPs bring about awareness and offer opportunities for organisations to gather supporters to help lead them to the next steps of the process, whether that's direct challenges (like boycotting plastic straws) or joining them on a peaceful march. Petitions work as an avenue to turn supportive 'armchair' conservationists into active participants in a movement. By signing a petition, you will be notified on its progress and get sent the link as soon as a debate goes live. Cristine adds, 'the most watched debates are the petitions' debates. Although things like Prime Minister's Questions *are the televised debates that people might recognise and say are the most watched, they're not.' If enough people are aware and engage with the issue, as petitions promote, then change will happen.*

Hope is the driver that underpins all these actions. Eighty-seven per cent of people are making the conscious effort to live a more sustainable lifestyle, and 72 per cent of them said they were motivated to do so for the sake of their children's futures. As individuals and communities, the decisions we make will have a lasting impact. Remember, you're in the driving seat and the corporations and big companies are your wheels. If you turn right, they will turn right too. Eventually.

HOW PLEASANT IS OUR GREEN?

We currently live in the New Forest National Park, an area that I have known and loved for over half a century. I consider myself very fortunate to be here; it's undeniably beautiful, and in places almost still peaceful. The ancient woodlands are home to rare birds, beetles and fungi, the open heaths have exotic plants and endangered invertebrates, and when the autumn mists settle over its mires and the sun rises and grasses glow it looks so picturesque. It is furnished with all sorts of designations to signify that it is ecologically important and it's recognised as a national treasure. That's why people visit in their millions each year to camp, cycle, ramble, go for cream teas and photograph the landscape and the animals that live here. But for me there's a problem: memory. I have a good one, which means that I can recall thousands of hours over thousands of days spent in thousands of places in the New Forest going all the way back until the mid-1960s. And it's just not as good as it was, nowhere near. There are far fewer birds, insects, plants and fungi, many of the 'good spots for a ...' are now empty, no longer any good and the habitats are in many places ruined. I see this, I have data – albeit limited to my own experience – but the visitors who come don't know what's missing, what's been destroyed. They see an often poorly managed mess dressed up as a postcard view of paradise. As I cycle past smiling holidaymakers snapping selfies I feel like saying, 'That barren billiard table of overgrazed and burned heath used to be alive with nightjars and Dartford warblers', or, 'That black muddy pit full of pony shit was once the verdant pool where I caught my first grass snake.' I don't, but as I pedal away I curse the ineffective protection that has allowed my favourite place on Earth to be despoiled. But how can this be? It's a national park isn't it? Shouldn't that mean that it's properly protected?

In terms of what protects, or is meant to protect, the UK countryside, there are a wealth of designations: Ramsar Site, Special Area of Conservation, Natura 2000, National Nature Reserves, Local Nature Reserves, Areas of Outstanding Natural Beauty ... the list goes on. But the most well-known are probably Sites of Special Scientific Interest (SSSIs) and our national parks. And I am afraid there is a widespread tendency among not only the wider public, but also those with a keen interest in wildlife, to over-rely on these as sanctuaries, as places where that wildlife can prosper and where we assume it is well-managed and cared for. So, let's have a quick look and get a more accurate assessment of what's really going on with our national parks.

Firstly, what is a national park meant to be? Well, their aims and purposes are quite clearly set out, although they differ slightly between England and Wales, and Scotland. For those in England and Wales, the Environment Act of 1995 revised the original legislation and set out two statutory purposes for national parks: 'to conserve and enhance the natural beauty, wildlife and cultural heritage' and 'to promote opportunities for the understanding and enjoyment of the special qualities by the public'. Scotland is slightly different, where there are four aims: 'to conserve and enhance the natural and cultural heritage of the area, to promote sustainable use of the natural resources of the area, to promote understanding and enjoyment (including enjoyment in the form of recreation) of the special qualities of the area by the public and to promote sustainable economic and social development of the area's communities'. The question is, are our parks actually meeting these criteria?

I think we could argue that some are not. Some, such as the Cairngorms National Park in Scotland and the North York Moors, Yorkshire Dales and Peak District National Parks in England, could be accused of failing to 'conserve and enhance the natural beauty, wildlife and cultural heritage' or to 'conserve and enhance the natural and cultural heritage of the area' by utterly failing to prevent escalating levels of wildlife crime, particularly the persecution of our birds of prey – the raptors. It could be argued that they are thus equally failing to 'promote opportunities for the understanding and enjoyment of the special qualities of national parks by the public' by failing to adequately protect these species and the environments where they live – the very things the public visit to see and enjoy. Significant areas within this suite of parks are given over to driven grouse shooting, where killing raptors, eradicating all other predators, burning moorland, facilitating the establishment of hill tracks, depositing lead shot in the soil and allowing the dispersal of medicated grit with unknown consequences for human and environmental health are commonplace, and certainly not compatible with promoting the 'sustainable use of the natural resources of the area' – they are irrefutably damaging it. By allowing a significant amount of this practice to govern the social and economic interest of these areas, to the detriment of other more inclusive and financially rewarding opportunities, they could also be accused of failing to 'promote sustainable economic and social development of the area's communities'. Indeed, these and some other national parks also seem to be in direct contravention of the Sandford Principle, a laudable recommendation first aired in 1974 when the National Parks Policy Review Committee was chaired by Lord Sandford. It states clearly that, 'If there is a conflict between protecting the environment and people enjoying the environment, that can't be resolved by management, then protecting the environment is more important.' This means protecting birds of prey from being illegally killed is more important than the activities of those who enjoy using the environment to legally kill red grouse. And to date there has been little or no evidence of a resolve to manage these peoples' activities to ensure the safety of that environment and its species, and therefore it seems the integrity of those national parks' underlying principles is compromised.

I know what you might be thinking: 'how dare he insult the crown jewels of our natural heritage?', but we shouldn't just sit back and consider these areas as sacrosanct and beyond criticism. I don't think we can afford to be so complacent and I wouldn't be the first to recently question the state or status of these swathes of green, but not so pleasant land. Kevin Cox, Chairman of the RSPB, raised the issue in March 2019, when an RSPB study found that on average only 26 per cent of SSSIs within English national parks are in a favourable condition in terms of their habitats and wildlife. He criticised the 'dismal performance' of these protected zones: 'SSSIs protect some of our most nationally important wildlife, and you would expect these sites to be thriving within national parks. The fact that these sites are, on average, in worse condition inside national parks than outside is therefore extremely concerning.' And the focus of his concern was that in England's national parks more than a quarter of the land – yes a quarter – is designated as SSSIs, meaning that it should provide refuge for the country's most precious wildlife. It stands to reason that non-designated habitats elsewhere within the parks could be in even worse condition ... who knows?

In real terms what this means is that failure to protect national parks is having a devastating effect on birds of significant conservation concern, such as lapwing, dunlin, common snipe and curlew, whose numbers are declining faster in the parks than elsewhere. In Exmoor National Park, curlew and kestrel numbers have collapsed, while the ring ouzel has become locally extinct despite the park until recently being a southern stronghold for the species. Tellingly, Cox's and others' beef wasn't with driven grouse shooting but with farming practices, as the intensification of agriculture has also negatively impacted biodiversity. In the Peak District National Park, for example, the number of sheep has increased fivefold since 1900, which has resulted in overgrazing, and thus a loss of vegetation, an increase in soil erosion and serious flooding incidents downstream. And overgrazing is not limited to this park – seven out of ten of our national parks are upland, where sheep stocking densities are an issue, and where I sit writing – in the New

Forest National Park – there are increasing concerns about the huge growth in the number of 'iconic' ponies and cattle. Most botanists I know are using another word to describe them!

THE ULTIMATE ARMS RACE

As the sun sets on a warm summer's evening, you look up to see small shadows skimming the air above your head. Just like shooting stars, they flash by sporadically, darting around as the cover of darkness falls and they begin a night of exploration and hunting. In the UK, we have eighteen species of bat, which account for over a quarter of all our mammal species. Globally, there are at least 1,400 bat species that represent about 20 per cent of all the mammals existing today. With the use of echolocation they are formidable predators, manoeuvring themselves with perfect precision to accurately catch flying insects on the wing. All UK bats are insectivores and have pretty big appetites to maintain their high-energy lifestyles. The common pipistrelle will eat over 3,000 flying insects in one night. Moths are a particular favourite, but they have come up with a cunning tactic to hide themselves from detection.

Many moth species evolved ears so that they could hear any incoming bat sonar and outmanoeuvre the predator, but one group, which includes the tiger moths, have gone a step further. When an echolocating bat gets too close for comfort, the moths emit their own ultrasonic clicks that jam the bats' sonar, disrupting their ability to gauge distance and confusing the picture of their surroundings. Research in 2013 found that this adaptation to avoid predation was having some other interesting benefits. When male Asian corn borer moths emit these ultrasonic clicks, the females are unable to distinguish between them and the real threat of an oncoming bat sending out sonar. In response to both bat and male moth clicks, the female freezes, giving the male moths an opportunity to mate with her. This is called a deceptive courtship song as the males take advantage of the female's predator response ... those sneaky little male moths! What a piece of evolution.

So far everyone is still buying the postcards and tea towels, having the cream teas and picnics, grabbing selfies with a view and loving the camping, but could these perceived paradises be in any real danger of losing their prestigious 'National Park' status? Shockingly, yes ...

The International Union for Conservation of Nature (IUCN) has a role in assessing our parks, and in particular their UK body, the National Committee for the UK. They publish a Statement of Compliance for the UK's national parks, which basically defines what the parks need to do to retain IUCN Protected Area status (Cat V – the big deal) and outlines each of the National Park Management Strategies. All are intrinsically linked to clear commitments to protect nature, to specific targets to achieve this and make sure that it is a priority. I hate to be really boring but in 'Annex II Assessment of National Parks Against Protected Area Criteria' it clearly states that 'For

National Parks in all three countries (England, Wales and Scotland), the conservation and enhancement of natural beauty, wildlife and cultural heritage is a first purpose'. And, if you'll excuse me continuing to be boring, critically in Annex III it asks, 'Does the site conserve the composition, structure, function and evolutionary potential of biodiversity?' All three nations controversially award themselves a YES. It also asks, 'Does the site maintain diversity of landscape or habitat and of associated species and ecosystems?' For all three nations the answer given is YES. It further asks, 'Does the site operate under the guidance of a management plan, and a monitoring and evaluation programme that supports adaptive management?' For all three nations the answer given is YES. And finally it asks, 'Does the site maintain the values for which it was assigned in perpetuity?' For all three nations the answer given is YES.

There are eight criteria and to retain its status a national park must score more than five – quite a perilous proposition given what I've outlined above. And the parks I've referred to above (Cairngorms, North York Moors, Yorkshire Dales and Peak District) could fail to meet these criteria purely given the levels of wildlife crime they tolerate, let alone their failings in other areas. But let's now ask practically what can these national parks do about it, and what do they want to? Probably not much ...

You see, they are powerless because they are governed by the regionally or locally powerful. Their boards are the problem – a familiar issue in the conservation sector – because they are made up of all the people who have an interest in their own interests not changing. Unless it's in *their* interest. Yes, all the driven grouse shooters, the farmers, the landowners are sat fat in those council chairs, which should be filled with independent ecologists. But even if this situation were reversed it wouldn't matter because of one colossal, catastrophic, but predictable flaw in the UK's national parks model ... They – don't – own – the – land. So ultimately they have precisely no control over what happens to or on it. Our national parks' boards are really nothing other than regional planning authorities. They can get uptight about housing and transport infrastructure (until they are overruled by the government or corrupted by their own corruptors), but they really have no sway over wildlife management, land use or abuse, and therefore conservation, so ultimately there is little point in picking on them – it's the system that is wrong. What a mess; what a sorry, predictably British mess.

Sadly there is another mess of a similar ilk, because despite what 'they' say there are real conflicts of interest between shooting and conservation in the UK. That's not to say that all shooting is bad, all shooters are bad, and none of their activities benefit conservation. It's not, they're not, and they may well do – but things like wildlife crime, the use of lead shot, the continued harvesting of endangered species and the ecological impact of releasing millions of non-native birds into our countryside can unfortunately overshadow any positive aspects of the activity. Central to this conflict is a lack of regulation.

You see, the UK has some of the most intensive game bird management anywhere in the world, but it's all very poorly regulated compared to other countries. Here game shooting is only controlled by having an open and closed season, which restricts the time of year when animals may be shot, and by firearms legislation, which places restrictions on who may have access to guns. There are some other laws governing the use of traps and snares, but these are rarely – if ever – enforced. This lack of regulation contrasts sharply with the licensing systems in place overseas. In Germany and Spain, for instance, there are effective powers to remove hunting licences and firearm certificates where wildlife crimes are committed, and there are strict habitat management plans in shooting areas. 'Game bags' measure the totals of each species that have been shot, locally, regionally and nationally. Elsewhere in the world game bag returns are required in order to inform real conservation for the populations of game birds and other species. In the UK they are all but non-existent, so any broader claims that shooting is good for conservation are easily undone. How can you conserve something if you don't even know how many of it you've been killing? UK game shooting has evolved without enough questions being raised and checks in place to limit its potential harm. This makes reforming it harder, because it's so much easier to not 'give' in the first place than to 'take' away later, especially when factions of that particular fraternity seem so adamantly opposed to change. So what should we do?

Well, in line with other nations we could introduce licensing for both shooting estates and individual shooters. They could, again in line with other European countries, include a two-part practical and theoretical examination to ensure the suitability and competence of the applicants. We could also implement the ability for regulators to permanently revoke those licences for an estate or individual if the law is broken. And surely we have to urgently introduce strict harvesting quotas and independently scrutinised bag monitoring to properly understand the impacts of shooting and inform conservation. When it comes to licensing, we would have to decide who pays for its origination, implementation and management – I would suggest surely the shooters themselves. Er, that idea probably won't receive a very warm welcome from the shooters as you and I currently subsidise the UK shotgun licensing scheme. But we don't do that for driving licences or passports, so why a gun licence?

The current cost of a shotgun licence renewal is £49, but police forces say the actual administration cost is in excess of £200, meaning that they and we taxpayers shoulder the burden. Further, as part of the application process our beleaguered NHS GPs are required to supply information about patients seeking a licence, but are not paid to do so. I can't believe that in the time of Covid-19, and all the pressures it is generating in the healthcare sector, the NHS should be subsidising non-NHS work. Of course, it is very obviously important that relevant medical conditions be flagged to police, but there is no reason why – in line with applications for pilots, divers, parachutists and other private hobbyists – shooters should not also be fully charged for their medicals. What makes them so special?

But does this lack of regulation in UK shooting have any actual effects on our resident wild bird fauna? Are there any negative impacts as a result of shooting? Again, this will probably strike you as absurd, but every year UK shooters legally kill birds of conservation concern while the rest of us lament their declines and do everything we can to protect them. Woodcock and common snipe can both be legally shot in the UK despite declines in their UK populations between 1974 and

1999 of 76 per cent and 67 per cent respectively. Woodcock are red-listed and snipe are amber-listed. The reasons for their decline include habitat loss and drainage, and not solely shooting, and woodcock shooters claim that shooting after 1 December avoids killing UK breeding birds. But the shooters' own data confirm that 17 per cent killed are indeed resident birds and, besides, the migrant populations may also be declining. I'm told an increasing number of responsible shoots have voluntarily ceased hunting these species, but the numbers shot are at a historically high level. The truth is that the effects of this harvesting on these species' populations are unstudied and unknown. That's why I think we should implement an immediate moratorium on the shooting of these birds, and pochard and golden plover, until rigorous and independent research tells us what's really going on. Isn't that common sense?

I'd say so, but then as any basically-educated ecologist would tell you, the effects of introducing a minimum of 43 million non-native ring-necked pheasants and nine million red-legged partridges into the UK countryside each year is also very likely to be having an impact on our native flora and fauna, and thus should also be better understood through similar rigorous and independent research. These vast numbers are released to be shot, presumably because native species such as grey partridge, black grouse, woodcock, etc have all but vanished. But in line with the overall lack of regulation in UK shooting we don't actually know exactly how many of these birds are released to be shot, nor what impact they have on the ecology of our countryside. What we all know is that the release of other non-native species is strictly controlled or illegal, and that given the available – but incomplete – data, we can estimate that half the biomass of our British birds in late summer is made up of pheasants. Yes, half the living weight of birds at large across our countryside is made up of non-native animals that are only there to be shot. It sounds crazy, doesn't it? Especially when they are implicated in the declines of snakes and lizards, they have been suggested as a reason for declines in woodland butterfly numbers, and their presence may deplete the food resources of other seed-eating birds simply because they are so numerous and are able to displace smaller species from natural food sources. Their super-abundance may also artificially increase the populations of generalist predators such as foxes and carrion crows, which in turn may have a disproportionate predatory impact on rare native species. And to ensure enough of them survive to actually be shot, an unknown quantity – but not fewer than hundreds of thousands – of native mammal and bird predators are legally killed each year. The crowning lunacy is that a recent shooting industry report has candidly revealed that only 38 per cent of those game birds released actually end up being shot.

HAVE YOU EVER RACED A T-REX?

When I talk about the world's fastest animal, I am sure that for some a cheetah will spring to mind. These long-legged cats can sprint at short distances up to about 70 to 75 miles per hour; it's fast, but they don't take the title. Rocketing through our countryside is a bird that can reach up to

200 miles per hour when stooping in good conditions – it is, of course, the peregrine falcon.

How some animals are able to move so much faster than others is a puzzle that has been confusing scientists for years, but new research has hinted at an answer that's all to do with size. Logical thinking would suggest that an animal's speed would have a strong relationship with limb or fin length, because its speed would impact how that species interacts with its environment – i.e. migration or, perhaps most importantly, whether it is predator or prey. But, as the fastest animals are not the elephants or whales of the world and are instead the cheetahs, sail fish and falcons, the evidence points in another direction. Bigger animals have larger bones and muscles that would not be able to cope with the forces exerted during rapid, quick movements. At the German Centre of Integrative Biodiversity Research, Myriam Hirt examined 474 species that run, swim or fly. The fastest species are those with a body mass that lies within the middle range; species whose body mass is over that intermediate threshold have significantly reduced maximum speeds. A peregrine falcon weighs between 600 and 1,300 grams, depending on its sex, so between its weight, hunting strategy and consequent physical adaptations it has taken the top spot.

This model for testing speed against body mass can work for animals that have different forms of movements in a range of environments, although there is more research that needs to be done on the model. It could potentially lead to some interesting discoveries, like predicting just how fast a *T. rex* could run or how nippy a megalodon shark was. However, research in this field has already suggested that most humans could have out-run the iconic *T. rex* so, based on biological accuracies, *Jurassic Park* would have been a very different story ...

One winter's day in the mid-1970s, while cycling to Hursley to look for a kestrel's nest in a quarry, a car passed me and clipped a cocky cock pheasant that was parading its finery in the road. A puff of copper plumes, beads of ruby blood and an instantly limp body. The car didn't stop, so I wrapped up the bird and carried it home in my ARP bag. My mother wasn't keen, so I had to pluck and dress it (I made a mess of both), but she cooked it and I ate it. I think I remember it being nice, I certainly had to say it was given the palaver, and I wondered why everyone was eating battery chickens when there were so many pheasants being blasted every winter. Nowadays an unknown quantity of shot pheasants are wastefully dumped because, as a recent review by Savills World Research for the shooting industry reveals, the market is so saturated they have little or no financial value. It's a sorry tale of waste, and of wasted lives.

It reports that 'during the 2017/18 season many shoots began to find marketing shot game increasingly challenging. The increased popularity of game shooting has led to more birds being released, creating an imbalance between game meat supply and demand'. It seems that game dealers only take 48 per cent on average of shot game and further that prices received have fallen by 50 to 60 per cent over the last six years. The same report details that £34.61 was the average outlay per bird before it was shot and yet not all shoots were able to sell their shot game. Last season a remarkable 46 per cent were supplying their game dealer free of charge and 12 per cent were actually paying them to just take them away at the cost of 20 to 30 pence per bird. No wonder they dig pits and dump the carcasses ...

The profitability of this enterprise is also interesting. In 2013/14 just over half of shoots made a loss. In 2017/18 they fared better, and the proportion was down to 42 per cent and collectively they made an average profit of 8 pence per bird, which is better than the 2016/17 season when the average shoot lost 36 pence per bird shot. Of course, it doesn't help that they are killed with lead shot, meaning that consuming them represents a significant public health risk.

In the first century Dioscorides, a physician in Roman emperor Nero's army, remarked, 'Lead makes the mind give way'. We've known that lead is poisonous for a very, very long time. Even low levels of lead are toxic to humans and therefore lead was banned from use in petrol, paint and water pipes in the UK years ago. Why then does lead ammunition remain in use? At least 5,000 tonnes of lead shot and bullets are blasted into the UK countryside every year, and 19,000 tonnes into European soil. And it remains there, a toxic legacy that continues to kill a large number of birds it wasn't even fired at. Ducks, swans and geese eat spent lead gunshot directly, mistaking it for grit, which they use in their gizzard to aid digestion, and predators and scavengers ingest shot, or fragments of it, when they eat dead game animals. When it enters the body lead paralyses muscles, modifies behaviour and impacts negatively upon reproduction, and when enough is absorbed it kills them. One piece of lead shot is enough to kill a small water bird. Work by

World Wildlife Trust (WWT) scientists estimated that 50–100,000 wildfowl die of lead poisoning each winter in the UK, along with many more terrestrial birds. The figure for Europe as a whole is a staggering: one million birds dead because of lead. So while we scrabble to acquire, manage or maintain all our wonderful wetland nature reserves and raise our binoculars on cold misty mornings to marvel at those swirling flocks, many of which have travelled across continents and oceans to visit us, they are out there in marshes and on the mudflats dying en masse because of a lack of responsibility and regulation from the UK's shooting fraternity. It beggars belief.

And that's just the birds, because as well as polluting the environment, when lead shot burns into the body of a dying animal it fragments and leaves tiny, often invisible, particles scattered throughout the animal's tissues. These fragments are then consumed by people eating the game meat, putting them at risk, especially children and pregnant women. While such exposure is very unlikely to kill humans, thousands to tens of thousands of children in the UK every year are estimated to eat enough game shot with lead to permanently reduce their IQ by at least one point. Which might go a way to explaining the notions manifested by some of my Twitter trolls!

But worst of all, this poison raining down over our landscape and into the bodies of our unsuspecting wildlife is completely unnecessary. Perfectly usable alternatives to lead ammunition have existed for years. Denmark banned the use of lead gunshot for all shooting way back in 1996, Norway in 2005, in Germany it's banned in shotguns, in New Zealand for twelve-bores and in the US it's banned for wildfowl shooting and, in some states, upland hunting too. Things have begun to move in Europe as well, where the European Chemicals Agency proposes to restrict the use of lead shot in 'terrains other than wetlands' by 2022. They have been investigating the issue and the call for evidence expired in December 2019. The report is expected in October this year and it's likely that it will finally lead to the EU issuing a blanket ban on the use of lead for hunting. So how is UK shooting dealing with this? Well, it seems to be worried, as they are about any mandatory regulation, so in February 2020 nine of their membership bodies voted not to support a ban on lead shot, but to sign a statement in support of the removal of lead and single-use plastic from all ammunition used for hunting game within five years. So that's another 25,000 tonnes of lead fired into our environment under a voluntary agreement. Diggory Hadoke fessed up in his March 2020 *Shooting UK* piece, in which he said, 'Hopefully, this will knee-cap those currently driving the legislative agenda and persuade the Government to take a pause and see how self-regulation changes the situation, reducing the motivation to find parliamentary time to create a legal ban on lead ammunition.' So it's another ruse then, to keep things the same? A much better and more pragmatic approach was taken by *Fieldsports* magazine, who recently published an essay by Thorkild Ellerbaek, a Danish shooter, about his experiences of the enforced switch from lead to steel shot and other aspects of regulation. In it he thoroughly lays to rest all the myths that afflict the intransigent, maudlin, luddite refuseniks who pepper the UK shooting press – in short, he says, 'get over it and get on with it'.

Enough of the ills of shooting; they're serious, but in the great grand scheme of things that harm the UK's wildlife they are dwarfed by those that arise through some of our farming activities. And that is why there is a growing concern among conservationists, particularly in regards to intensive farming. In 2017 a scientific report revealed that 76 per cent of flying insects had 'vanished' from German nature reserves over the last 25 years. The following year two studies in France recorded a decline of 30–80 per cent in farmland bird numbers in the last fifteen years, matching our own UK figure of a 56 per cent reduction between 1970 and 2015. In all of these cases habitat destruction and pesticide use are implicated as the primary reasons for the declines. Bob Dylan famously sang, 'You don't need a weatherman to know which way the wind blows', and you don't need me to tell you that if these trends continue then we are facing an ecological apocalypse.

The State of Nature Report 2016 revealed that between 1970 and 2013, 56 per cent of UK species declined. Of the nearly 8,000 species assessed using modern criteria 15 per cent are threatened with extinction, which suggests that we are among the most nature-depleted countries in the world. Of the 218 countries assessed for 'biodiversity intactness', the UK is currently ranked 189, a consequence of centuries of industrialisation, urbanisation and overexploitation of our natural resources. Like I said, we are facing an ecological apocalypse. And the reason I've said it twice is that somehow we have grown to accept this as part of our lives – we've normalised the destruction of our wildlife. And worse, to our shame, we are even careless with our language. Standing there moaning in nature reserve car parks, we say that 'we've *lost* 97 per cent of our flower-rich meadows since the 1930s' or that 'we've *lost* 86 per cent of the corn bunting population'. We grumble about 'a *loss* of 97 per cent of our hedgehogs'. Loss, lost ... like this habitat and these species have mysteriously disappeared into the ether, as if they've annoyingly, accidentally vanished. They haven't – they've been ploughed up, they are dead, or they don't exist. *Destroyed* is the word we should be using. And much and many of them have been destroyed in our lifetimes, on our watch.

SUMMONING LIGHTNING

As a young girl I can remember lying among the blades of grass on the untidy lawn at the front of the house with my beady little eye peering down into the burrow of a stag beetle. It was emerging slowly, mandibles first, and I was in awe. What an animal.

Named after their oversized mandibles, which resemble antlers, these impressive beetles are one of the UK's most spectacular insects. Depending on the weather, stag beetles spend somewhere between three and seven years underground as larvae, feasting on decaying wood and returning important minerals back into the ground, before building a cocoon in the soil, pupating and metamorphosing into the adults we see fondly above the ground. They exit their burrows in mid-May to breed. You can spot them sunning themselves during the day and flying, albeit slightly clumsily, at dusk in search of the perfect mate, but sadly they'll only survive that summer.

My child-self found it astonishing – well I suppose I still do – how an animal could live underground for years gathering the energy to transform itself into an armoured adult beauty, just to perish within weeks. I suppose it is just a matter of perspective as that stag beetle larvae occupied a narrow niche and contributed to the ecological integrity of the UK's woodlands through the decomposition of fallen trees. With numbers declining due to the lack of deadwood habitat that they heavily depend on, we must remember that the logs and wooden stumps in our gardens and local green spaces may seem useless to us, but to a whole host of biodiversity it is their life-support system. And in turn, we depend on it also.

It is the folklore surrounding the stag beetle that I find simply fascinating and, as with any old mythologies, a species deemed bad in one country is often celebrated by another. In the UK during medieval times many believed that these beetles were associated with the devil, as they emerged from below the earth with their reddish-tinged 'horns'. In the New Forest, they were nicknamed the devil's imps and on their arrival from the soil they would be stoned out of the air because the local community associated their arrival with a bad corn harvest. In other areas around the country, stag beetles were blamed for summoning lightning and causing fires by carrying hot coals in their mandibles, dropping them onto the roofs of houses. But it wasn't all bad news for the stag beetles elsewhere in Europe. In northern France, carrying a stag beetle would supposedly bring you great wealth and in Romania these beetles would ward off the devil's eye.

There is obviously no doubt that industrial farming is a central part of the problem, but we do have to be very clear that farmers can and need to be the most effective part of the solution. Just as there are good shooters there are very, very good farmers; it's simply that the number who are farming in harmony with wildlife and the area they are improving is still way too small. The excellent Nature Friendly Farming Network describes this cohort as 'many', a term often quoted widely in the farming press, but its subjectivity simply hides the fact that this 'many' are not yet contributing anything meaningful beyond their local patch, because there are just not enough of them. Because just like other conservationists theirs is a movement motivated by personal energies. The broader farming sector is not being properly encouraged to join in and one of the principal barriers to this is the National Farmers' Union (NFU).

This organisation is neither national nor representative of all farmers' interests, and nor is it really a 'union', as in a democratic association of workers created to help represent their collective interests in negotiations with their employer. As highlighted by the Ethical Consumer investigation into the premise and practices of the NFU, 'English Agribusiness Lobby' would be a better name. Scotland, Wales and Northern Ireland have their own unions. The attention of these 'unions' to the interests of smaller farmers is slight compared to the attention given by them to larger intensive-farming methods, and their relationships with powerful agrichemical companies such as Syngenta are notable and significant. These 'unions' don't appear to like science much, unless it suits their agenda; in the teeth of the weight of scientific opinion they have been keen advocates of the badger cull, they steadfastly fought against the withdrawal of the neonicotinoid pesticides and continue to resist restrictions in the use of glyphosate.

Sadly the NFU don't like conservationists much either, doing little to encourage relationships between us and farmers; indeed some of their members have branded us 'anti-farming', thereby polarising the two obviously closely allied groups. Sadly, this has found traction in the farming fraternity, especially in the large chemically dependent and intensive sector. This is disappointing and especially harmful when the wholesale declines in biodiversity due to intensive agriculture must be addressed by farmers and conservationists together. It's a problem, but one that could be solved by exposing the actual agenda of the farming 'unions', restricting their lobbying power within government, encouraging them to embrace a real interest in wildlife-friendly farming initiatives, including a properly proportional representation and promotion of organic farming, and educating their members to implement clear science-led policies and more sustainable long-term farming strategies. This means reducing our dependence on intensive farming and the things that fuel it – pesticides and fertilisers, the ills of which we've known about for a long time. We were all warned of the possibility of a silent spring in way back in 1962, but the regulatory system governing pesticide use has repeatedly failed to prevent harmful chemicals from being approved for use in our fields. Organochlorides, organophosphates and, more recently, neonicotinoids were only banned after decades of use and enormous environmental damage. You see, it may look green and pleasant as you walk, drive or rail through it, but modern industrial farming sees the repeated application of multiple pesticides to our landscape on a breath-taking scale. As Professor Dave Goulson from the University of Sussex wrote in my 2018 initiative *A People's Manifesto for Wildlife*, 'About 500 different "active ingredients" (i.e. poisons) are licensed for use in the EU. In 2016, 16.9 thousand tonnes of "active ingredient" were applied to the farmlands of Great Britain, comprising 5.9 thousand tonnes of fungicide, 7.8 thousand tonnes of herbicide, and 315 tonnes of insecticide. Pesticide use continues to rise; on average in 2016, each farmers' field was treated with seventeen different pesticide applications, approximately double the number 25 years ago. In short, our farmland is being subjected to a massive barrage of poisons, leading to contamination of soils, hedgerows, rivers and ponds. All farmland wildlife is being chronically exposed to a complex mixture of pesticides, the effects of which are far beyond the capacity of scientists to predict or understand'. So why are we doing this?

Well, we are so frequently told that pesticides are absolutely essential if we are to feed ourselves that we've come to believe it. But recent studies actually suggest that a lot pesticide use is unnecessary and that most farmers would save a lot of money if they used fewer of these compounds. Unbelievably, many pesticides are now sprayed onto crops just in case a pest appears, rather than in direct response to it being there causing damage. And as Prof Goulson says, 'despite the enormous number of pesticides and synthetic fertilisers used in industrial farming, organic farming manages to produce on average 92 per cent of the yield of the same crops. And yet organic farming has had almost no investment or research and if it did then it's highly likely that this 8 per cent gap could be closed even further'. In stark contrast, billions have been invested in developing new chemicals.

France and Denmark have recently set clear pesticide use reduction targets of 50 per cent and 40 per cent respectively. Denmark has also introduced a pesticide tax representing 34 to 55 per cent of the sale price of the chemicals. Maybe the UK should do the same and use the revenue from the tax to fund an independent advisory service for farmers to test the effectiveness of pesticide reduction measures? Maybe we could set more ambitious targets to ensure that a greater proportion of the UK's farmland would be organic and introduce a series of bans? You know, not commission reports or reviews and hold lengthy consultations with people like the NFU, Bayer or Syngenta, but actually just stop doing the things that are doing us and our wildlife most harm. Now there's an idea!

So, let's ban all non-agricultural uses of pesticides for use in parks, and in national parks too. Let's ban pesticide use in our cities; Vancouver has done it and an increasing number of UK boroughs have restrictions in place. And, sorry, let's ban the sale of pesticides for use in gardens. And finally let's ban glyphosate, the principal ingredient in the weed-killer marketed as Roundup, with a time-limited withdrawal from use to allow for alternative weed control methods to be developed for use in 'no-till' farming systems. These systems are better for retaining good soil structure and moisture, reducing soil compaction and erosion and cutting labour and fuel costs. But while they are thus more environmentally friendly, any advantages are destroyed when the fields are drenched in a chemical agent that, according to the World Health Organization's International Agency for Research, 'probably' causes cancer. And it's everywhere; described as the world's most popular herbicide, glyphosate's worldwide use has increased almost fifteenfold since the mid-nineties when genetically engineered 'Roundup Ready' crops were introduced. In the UK it is one of farming's most popular weed-killing brands – according to a 2017 study of government data 5.4 million acres of farmland across Britain are drenched with glyphosate every year.

Roundup was made by Monsanto, a company that was bought out by the German chemical giant Bayer in 2018. This acquisition has proved a bit tricky for them and owing to ongoing litigation concerning Roundup the takeover is considered one of the worst corporate deals ever agreed – it has caused Bayer massive financial and reputational blows. Glyphosate is obviously a poison – it kills plants, you'll see that when you look over those dead orange fields that stretch across our countryside – the question is, how poisonous is it to humans and wildlife?

Just two months after Bayer acquired Monsanto a US jury ordered Monsanto to pay $289 million to a school groundskeeper who claimed his non-Hodgkin's lymphoma was caused by regularly using Roundup. Following that verdict Bayer's share price plummeted by around 14 per cent, which cost the company $14 billion in market capitalisation. Bayer predictably filed an appeal, but then in 2019 another court ordered Bayer to pay more than $2.5 billion in damages to a couple in California who had both contracted non-Hodgkin's lymphoma. But it gets worse ... in June 2020 the company agreed to pay a staggering $9.6 billion to settle more than 10,000 lawsuits claiming harm from Roundup. According to the company, this doesn't mean that they admit liability or wrongdoing, so they are not admitting that Roundup, essentially glyphosate, is a human carcinogen, but they have set aside $9.6 billion to compensate those who claim it is. That's very odd. In fact what's downright strange is that the general consensus among national regulatory agencies, including the European Commission, is that usage of the herbicide when the label's directions are followed poses no carcinogenic or genotoxic risk to humans. And in January 2020, the United States Environmental Protection Agency finalised its interim registration review for Roundup and stated that it 'did not identify any risks of concern' for cancer and other risks to humans from glyphosate exposure.

In 2019 Bayer's total assets were worth an eyewatering €126,258 billion– these are the big boys, a global megacompany with unlimited funds to lobby the people in power. I've seen that in action too. In 2017 I was visiting the European Commission to present evidence seeking to reform the unsustainable hunting of turtle doves on Malta. As we made our way to the chamber we passed through hordes of smartly suited people hovering in huddles along the concourse. The Dutch MEP who was our host told me these were all 'glyphosate lobbyists', poised and ready to pounce on their targets as they emerged from their meetings and debates. There were droves of them, and they won. The Commission granted a licence for a further five years and glyphosate is currently approved for use in the European Union until 15 December 2022. The UK was among the states in favour of glyphosate renewal, and although they had previously abstained, Germany and Poland were also among those who sought to prolong its use. France and Belgium voted against and Portugal abstained, later banning its use in all public spaces. It seems the writing is on the wall as Germany has now announced that it will begin phasing out the controversial weed-killer by 2023 and by the end of 2020 glyphosate will be totally banned in the Grand Duchy of Luxembourg, making this the first EU country to screw the cap firmly shut on this herbicide. But when I began to question its use on social media the reaction from the agrichemically dependent industrial farming lobby was vicious and the NFU wrote to the BBC demanding I was sacked. On Twitter one advocate of the toxin was bragging about having drunk a pint of Roundup. I wonder if he is still alive?

But we must separate such lunacy from the UK farming fraternity, which desperately needs our support. Our constant hunger for the cheapest food means that someone out there is paying the real cost – our farmers. Many struggle to realise a profit on their produce, thus becoming dependent on our tax handouts, because we rush to supermarkets to buy cheaper food imported from overseas. New Zealand lamb: why the hell is this on our shelves? Shipped halfway round the world with an absurd carbon footprint to undermine the businesses of one of the most beleaguered groups of our farmers, who then need enormous support with those tax handouts. Why are we not encouraged to buy UK lamb? I'm not promoting nationalism, or even patriotism, here, it's just common sense isn't it? We must start putting our pounds into our farmers' pockets – especially our organic farmers' pockets – even if it costs us a little more. Otherwise how can we ask them to do this, that or the other for conservation if we turn our backs on their troubled economy in the aisles of Tesco, Sainsbury's or Waitrose? They are the only man and woman power out there on that 70 per cent of our landscape who can actively make the difference. So please support ethical, wildlife-friendly farmers, and help them lead the way to a new farming future. A future where wildlife thrives.

Nature Friendly Farming Network

The Nature Friendly Farming Network (NFFN) was mentioned briefly earlier on, but their work is so instrumental that we couldn't write a book about putting nature back together again without discussing them further. The NFFN showcases how farming can and should be done for the benefit of farmers, their yields and profit, in conjunction with promoting a better environment and richer biodiversity.

The UK's landscape is sculpted by farmland, with over 70 per cent of our land being given over to it. That's a big old chunk of land that is drained, sprayed and harvested, but the NFFN are doing things differently. They were formed by a coalition of passionate farmers who have seen first-hand the devastating declines of up to 600 farmland flora and fauna – including turtle dove (<89 per cent), corn bunting (<90 per cent) and tree sparrow (<94 per cent) – in the last 50 years. We've already spoken of the damage caused by intensive agriculture on our landscape, but who better to champion an alternative method than the farmers themselves who know their land, its ecology and the species that reside on it? If you think about it, farming was once a product of nature, until we over-managed it and blurred those lines, but now the NFFN are taking a big step forward by stepping back *and are leading the way to a new age of agriculture. Here are some of their stories.*

Martin Lines is the UK chair of the NFFN and is someone who I have had the pleasure of meeting on several occasions. Chris and I visited his farm in 2018 as part of the #WeWantWildlife bioblitz tour, where we visited 50 conservation sites in ten days to assess the state of nature in the UK. Thinking back on it, Papley Grove, a 900-acre arable farm run by Martin and his team, was one of our highlights as it brilliantly showcased a bright and exciting way forward to benefit all. Martin is a third-generation farmer who took over the farm from his father and has since focused on reconnecting nature with farming. I am no farmer, so don't just take it from me – Martin says, 'For over ten years our farm was in the old Countryside Stewardship Scheme to try to improve the natural habitat for wildlife on the farm. We restored many of the hedges around the fields that had previously been removed, improving the few that were remaining and planting new ones. We also established grass strips alongside hedges and ditches, and on our field boundaries. Over this time, we saw a significant increase in wildlife, both flora and fauna. The RSPB undertook several surveys, which identified the wide range of species found on the farm, including birds of high conservation concern such as turtle doves, yellow wagtails and corn buntings. In the future, I hope to continue and extend our conservation work and link up wildlife habitats on neighbouring farms. We have made many wildlife corridors across the farm to help the wildlife move about. I continue to run a productive arable business, alongside a great wildlife package.'

Stepping out of the car and into Martin's farm on our visit a couple of years ago felt very different. I have been on many farms, but none quite like this one, which was brimming with life. The hours we spent there were filled with searching for damselflies, butterflies, birds and small mammals. Over a 24-hour period, Martin and the team of wildlife recorders found 329 different species. I remember leaving feeling inspired by his mindset and determined to make a difference.

When Neil Haseltine, a nature-friendly farmer at Hill Top Farm in North Yorkshire, was asked what impact this approach had had on his farm, he responded, 'By reducing the stocking rate – switching to cattle grazing and changing the times of grazing – we have changed the environment dramatically. Botanically, it is very different. A lot of plants have returned, including rock rose, bird's eye primrose, scabious, wild thyme, spearmint, bluebells. We never used to see these species. We never used to have barn owls; now the RSPB ring chicks every year. I used to see a hare about once a year, now I see a hare one in every three days. There are also more skylarks, redshanks and curlews.' Over the last fourteen years, Neil has reduced the stock density on his 1,100-acre farm from 800 sheep to just 190 Swaledale sheep and a breeding herd of 30 Galloway cattle, which are breeds specifically suited to his upland farm environment. He added, 'Reduced stocking pressure has also allowed the grass to grow longer, which may help prevent flooding downstream. We have also put in several mechanisms for natural flood management. Reduced stocking density has had a positive impact on our profits. Overall output has decreased, income from agriculture has decreased, but we have become more profitable. This is because the costs of production are so much less. Fewer sheep means we don't need to buy in concentrates or feeds. We don't need extra people to help on the farm. We turned this farm from a loss-making to a profit-making enterprise. And this is before you take any support mechanisms into account.'

As environmentally conscious consumers, as we have already said, it is our duty to put our pounds and pennies where it matters most. Supermarkets are constantly pushing down prices and importing increasingly cheaper goods from overseas. Gethin Owen is another NFFN farmer in North Wales, and when posed the question 'what do you need from citizens – how can they help?', he replied, 'we need the consumer to create the demand for food produced using nature-friendly methods. We are already seeing large supermarkets influencing standards. The consumer needs to be sufficiently well-informed, have an interest in the provenance of food, and consider the consequences of the choices.' On a broader note, discussing the future of schemes and subsidies, Gethin said, 'historically, farming policy has encouraged production at all costs, and any funding for delivering environmental benefits has been minuscule in comparison. Policy has favoured the large, established farmer and discouraged the younger generation who want to do things differently. Giving money to farmers just because they farm needs to stop. Future schemes need to be outcome-based, rewarding farmers for enhancing biodiversity, improving water and air quality and capturing carbon. These schemes need to be for the actual person managing the land and not the landowner, and should incentivise the landowner to give security of tenure to the person who is actually managing the land.'

The importance of nature-friendly farming was poignantly summed up by Neil Haseltine, who said, 'We have chosen to focus on the natural and sustainable farming route on our farm. For us it is not only a more profitable way of farming, but also more sustainable, from an environmental and economic point of view. We are more likely to be still farming in ten, twenty, a hundred years' time if we don't try to work against nature. Unsustainable practices have a finite lifespan.'

Can't say truer than that.

WHERE EAGLES DAREN'T

On a bleak April day high in the Cairngorms National Park a policeman zips up his raincoat, tugs on a pair of blue rubber gloves and crouches down over a corpse. It's a grim and gruesome scene, but one he has become increasingly familiar with; the bodies have been piling up and although he, and everyone else, knows who the killers are, they've yet to face their day in court or any time behind bars. And that injustice is driving conservationists wild.

You see, the victim is a young white-tailed eagle and it's been found because it was fitted with a satellite tracking tag – a very sophisticated device that constantly transmits a plethora of biometric data – so those monitoring it didn't just know very precisely where it was, they also knew that it was dead. Full of a cocktail of illegal poisons on a driven grouse moor. A tragic end to a magnificent bird, an essential keystone component in the ecology of this area and a national natural treasure. Another victim of wildlife crime.

A report commissioned by the Scottish government and published in 2017, 'Analyses of the fates of satellite tagged Golden Eagles in Scotland', examined the outcomes for 131 eagles fitted with satellite tags between 2004 and 2016 and concluded that as many as 41 (31 per cent) had disappeared under suspicious circumstances, presumably having been killed. It stated that 'a relatively large number of the satellite-tagged golden eagles were probably killed, mostly on or near some grouse moors where there was also recent independent evidence of illegal persecution'. And this was just one species; as Ian Thomson, RSPB Scotland's head of investigations, pointed out, 'when you add the disappearances of satellite-tagged white-tailed eagles, red kites, goshawks, peregrines and hen harriers not included in this review, and consider that satellite-tagged birds form a very small proportion of the populations of these species, the overall numbers of eagles and other protected raptors that are actually being killed must be staggering.' The visible tip of an iceberg of death.

It's hard to imagine that in the early part of the twenty-first century in the UK – a place where enormous knowledge has been hewn from great ecological science to inform fantastic conservation, where more than a million people are members of one of the world's most respected ornithological NGOs, where the populace spends more than £350 million on feeding birds every year, in Britain, the 'nation of animal lovers'– criminals are still killing our birds of prey. Relentlessly, cruelly and daily. But it's true.

Does this crime have any real impact on the species' populations and distributions? Well, in the case of golden eagles the scientists who analysed the fates of the 'disappeared' satellite-tracked birds stated, 'This illegal killing had a marked effect on the survival rates of the young birds, so that we expect the potential capacity for the breeding golden eagle population continues to be suppressed in the environs where the killing largely appeared to occur (in parts of the central and eastern Highlands of Scotland). This is after decades of continued suppression through illegal killing in these same areas of Scotland.' In his excellent book, *Inglorious: Conflict in the Uplands*, Mark Avery cites another government-commissioned scientific report that had calculated that there is enough suitable habitat in England to support 330 pairs of hen harriers, but in most years they struggle to reach double figures. Again, persecution on grouse moors is to blame. A long-term Natural England study of tagged birds eventually published in 2019 revealed that hen harriers are ten times more likely to die or disappear on or near to English grouse moors than any other habitat. Analysis over a decade found 72 per cent of 58 satellite-tagged birds were confirmed or considered 'very likely' to have been illegally killed. Only 17 per cent of very lucky young hen harriers survived beyond their first year on grouse moors in northern England and southern Scotland, compared with 36 per cent across the whole Scottish mainland and up to 54 per cent on Orkney, where there is no driven grouse shooting.

Sadly it's not just hen harriers; in recent years just about every one of our resident species has picked up shot or been poisoned or trapped somewhere in the UK. The RSPB appear to be the only body validly collating this data, which is odd given the scale and nature of the crime. One would imagine that in England at least the Department for Environment, Food & Rural Affairs (DEFRA) would take a bit more interest, or maybe Natural England ... Of course, there is some posturing going on, and in 2011 the Raptor Persecution Priority Delivery

Group was formed. This group is police-led, as this issue has been recognised as a national wildlife crime priority, but the group includes all the usual suspects from the shooting fraternity, so while they have produced maps of where the dead birds have been found, the data has been significantly diluted through the meddling of interested groups, leaving the RSPB and others disappointed and disenfranchised. The way things are going, I can't see this project lasting as it has achieved little in terms of stopping the crime and has really only served as a stalling mechanism for those who protect the criminals.

And because killing birds of prey is a crime you may be wondering what the police and judiciary have been doing about it. In terms of action in the field, for a long time it was the RSPB's Investigations Department that did all the dirty groundwork. Tough physically and mentally – who would want to spend more time looking for and at dead beautiful birds than at beautiful live birds? This continues apace and for some time they have been working as closely as possible with a network of police wildlife crime officers. To be ruthlessly honest, at the start this wasn't functional – the officers were poorly trained in regard to wildlife crime and their work lacked support from senior ranks. When they were good they were great, but the NWCU hasn't always quite got the hang of genuine partnership working with conservationists. And on top of this, every time their budget came up for renewal the National Wildlife Crime Unit was threatened with closure. In 2016, after a significant public outcry, at the very last minute the government found £1.2 million to fund the department until at least 2020. Thankfully the recruitment, commitment and delivery from these officers has greatly improved and we now have police officers of all ranks with wildlife criminals in their sights. Operation Owl, a raptor crime reporting scheme, is going well in North Yorkshire and, away from raptor persecution, Operation Galileo, an anti-hare coursing initiative launched in 2019 by Lincolnshire Police, has seen a very significant reduction in this nefarious activity. Using drones and seizing dogs has seen incidents in the county fall by more than 50 per cent since 2016. A number of other forces are not yet performing optimally, an artefact of the personalities involved at that local level. The police can conduct inquiries, gather evidence, sometimes even make arrests but they, and we, are reliant on the judiciary to implement the full force of the law.

BEAT 'EM WITH SONG

There are many conflicts humans are responsible for creating that are unnecessary, unjust and unkind, especially as we should and do know better, but there are varying degrees of natural conflicts and competitions between species that are beneficial, driving evolution and behaviour. This is especially relevant to those species occupying the same niche, i.e. those in the same space at the same time, and with overlapping resource requirements. Bird feeders invite lots of biodiversity into our backyards, and we become witness to some of the best wildlife dramas around. You may remember that in *Springwatch* 2020 we spoke about how larger garden birds on the feeder tended to be more dominant, driving away the smaller birds.

Research shows that the house sparrows, greenfinches and great spotted woodpeckers rule our backyard buffets, while the coal tits and great tits place last. This is an example of interspecific competition (between species), but intraspecific competition is also a strong driving factor as individuals of the same species vie to win the best mates, food and habitats.

Common forms of intraspecific competition might be physical fighting, signalling through bright colouration or ornaments, or behavioural expressions via displays or vocalisations. For example, female and male blue tits are more likely to select a mate with plumage that has high UV-reflectivity. However, new research has also found some interesting conflicts between individuals when it comes to singing. The purpose of birdsong is either to attract a mate or to defend a territory, and while many birds sing primarily in the mating season, there are some that pump it out all year round. The Massey University in Auckland, New Zealand, found that male tui songbirds will be much more aggressive to other males trespassing on their territory if the trespassing individual is a good singer with a more complex song. In response, the original male will respond with a song that is longer and has more syllables to try to compete with the rival contender. This is the first evidence that song performance drives behaviour in birds.

The number of wildlife crime cases that make it to court is minimal. Recently in Scotland multiple cases where the evidence appeared to be strong went through up to a year's worth of preliminary hearings only to be thrown out on the eves of their court hearings. But some do get through and a few find the defendants guilty. Here is a tale of two Alans ...

In 2014 Allen Lambert, a gamekeeper on the Stody Estate in Norfolk, was convicted of killing ten buzzards and a sparrow-hawk, storing banned pesticides (used as poisons) and a firearms offence. Even though the judge acknowledged that his crimes had passed the custody threshold, he was given two suspended sentences (suspended for a year) and fined around £1,000. His employer did not sack him and he was allegedly allowed to take early retirement. But later the Stody Estate were hit with a massive financial penalty – the then-Environment Minister, Andrea Leadsom, withdrew 55 per cent of their farm subsidies, thought to be the largest ever civil penalty for raptor persecution crimes. Now That's What I Call Justice!

Except, it wasn't ... no, they appealed, the NFU intervened, it went to court and Mrs Justice May pronounced that, 'The mere fact of Mr Lambert's conviction did not prove that poisoning the birds was "directly attributable" to his employer.' The Stody Estate, which has been farmed by the MacNicol family for 75 years, were exonerated from any involvement in poisoning birds and the penalty was quashed. Think of that what you will.

More recently, Alan Wilson, a gamekeeper on the Longformacus Estate, achieved notoriety when in 2019 his 'kill lists' made the news. Goshawks, buzzards, badgers and an otter represented only the tip of the iceberg, and one such list found in Wilson's home catalogued 1,071 dead animals, including cats, foxes, stoats, weasels, rooks, jackdaws and hedgehogs. Most of the corpses went into a horrific 'stink pit', a disgusting but frequently used device designed to attract birds of prey and other animals, which Wilson was then suspected of shooting. He was found guilty of nine offences, including being in possession of 23 illegal snares and gin traps, and the banned pesticide carbofuran, which is hugely poisonous to animals and humans alike. So you can only imagine that the judge, in this case Sheriff Peter Paterson, would have thrown the book at him ... but no, Wilson was sentenced to just 225 hours of unpaid work. You couldn't make it up, and you don't have to, because the sheriff said this was 'a very severe case', but that he had no option within the confines of the law to hand down a custodial sentence due to guidance on short-term sentencing. Imagine how the diligent staff of the League Against Cruel Sports, the Scottish Society for Prevention of Cruelty to Animals and Police Scotland, who gathered the evidence and brought the case, felt when they heard that? Demoralised maybe, and probably, like the rest of us, incandescent with rage.

The wildlife criminals have been, and still are, getting away with it, thanks to the wilful blindness of politicians who just don't have the guts to get into this issue, the patchy effectiveness of police wildlife crime officers, the poor budgets given to the National Wildlife Crime Unit, the fact that it's too easy for these criminals to hide their crimes, the curious and unaccountable behaviour of the Scottish Crown Office and the Crown Prosecution Service and the inordinately weak sentencing from the courts. So, it's all 'their' fault. Or is it? No, I'd say in some ways it's our fault, because the vast majority of the public still don't perceive wildlife crime as real crime. It's not murder, rape, theft, fraud, it's not even despised with the same ferocity as drink driving. All of these are serious crimes, but until we think of the poisoning of that young white-tailed eagle in the

Cairngorms National Park as a similarly serious crime then none of 'them' will act to end it.

Public perception is rapidly changing, though, as has been displayed with an unprecedented response to an e-action set up in August 2020. Although brokered by Mark Avery and sanctioned by the RSPB, the credit has to go to Fabian Harrison, whose idea it was to ask people to send an email to their elected representative wherever they are in the UK. Following the annual Hen Harrier Day, a day of raising awareness in which conservationists highlight the plight of this iconic bird at the hands of wildlife criminals, we ran a campaign to get people to click on a box, enter their postcode and press send. The email was a very polite message simply asking that the recipient take more interest in this issue and understand its importance. I conducted a series of short social media interviews with change-makers and politicians, including Angela Rayner, the deputy leader of the Labour Party, and Caroline Lucas and Alison Johnstone from the English and Scottish Greens respectively. (We obviously invited someone from government – they declined to appear and sent a two-sentence statement instead.) Our discussions broadened the debate beyond hen harriers to include burning peat, animal welfare and the economic future of the communities where driven grouse shooting is currently practised. What was so very heartening and empowering is that within just three weeks no fewer than 123,701 people had participated. That huge number of emails were jamming politicians' inboxes and represent tangible proof not only of a burgeoning number of people who are sick of this and their determination to do something creative about it, but that we can all make a difference. The catalogue of ills outlined in this chapter can be depressing, but we cannot ever afford to get disheartened because as this simple-to-do e-action proved we are all empowered to drive change. And sometimes politicians manage to do the same.

Some raptors can eat grouse and pheasants, and that's why some gamekeepers kill them. But mountain hares, the UK's only indigenous lagomorph (both rabbit and brown hare have been introduced), are herbivores, so why have they been ruthlessly exterminated on some grouse moors? Well, it's thought that because they carry ticks and ticks can spread diseases, they might aid in the transmission of 'louping ill', a viral infection primarily of sheep, but which can also affect grouse. The link, and its possible importance, has not been scientifically validated. Indeed, in their 2010 *Journal of Applied Ecology* paper Harrison et al concluded that 'there is no compelling evidence base to suggest culling mountain hares might increase red grouse densities'. Nevertheless at least 26,000 hares a year are culled on Scottish grouse moors despite measured declines in their population. There is a closed season, but no licences are required and there are no inspections to confirm they are being humanely killed and no mandatory bag totals are kept to monitor the impact of the culling. There are claims that the hares enter the human food chain, complete with the lead shot that killed them, but often they are dumped to rot in huge piles and in gamekeepers' stink pits.

In life, the mountain hare has a huge asset: in winter they moult into a white pelage, which camouflages them in the snow but also makes them undeniably attractive, and thus their relentless persecution has led to widespread public antipathy that in turn has resulted in forceful lobbying of the Scottish government. Its retort, voiced by Cabinet Secretary for Environment, Climate Change and Land Reform, Roseanna Cunningham, was 'bring us the evidence and we'll do something about it'. And so animal welfare charities OneKind and the League Against Cruel Sports, and a film crew supported by the cosmetics company LUSH, went into the field and filmed the conspicuous and ugly slaughter of hares. And the government did act – in June 2020 the Scottish parliament agreed to make mountain hares a protected species under new legislation, the Animals and Wildlife (Penalties, Protections and Powers) (Scotland) Act 2020. This means mountain hares would only be permitted to be killed under licence, effectively ending the mass killing. Result? Well no, not yet. You see the wonderful Alison Johnstone, Green Party MSP, tabled the amendment to enact this legislation late, but not impermissibly late, in the process. It passed the vote, but other MSPs were miffed by this manoeuvring and Mairi Gougeon, Scottish National Party MSP, whose Angus constituency contains driven grouse moors, secured a process of consultation with stakeholders lasting two years before the licencing scheme is introduced. A timely delay because 1 August marks the start of the mountain hare shooting season, and the chances of the grouse moor keepers exercising any of the 'voluntary restraint' previously requested by Holyrood and ignored are negligible.

As with raptor persecution it's pertinent to ask if all this killing is having an impact on the hare's numbers. Harrison and her fellow researchers discovered that 'evidence from the available literature is limited', that 'this restricts our ability to reliably assess the effectiveness of culling mountain hares to control ticks and louping ill virus', and furthermore 'the population response of mountain hares to culling is not well understood and the possible effects on their conservation status and the upland ecosystem remain unexplored'. In short, we don't know, so in my book we should stop killing them at least until we do.

So far we've considered individual species that are illegally and legally harmed, but there are very many examples of not just species or groups of species being destroyed, but whole ecosystems being ravaged. One of the most prominent current cases of this is that of salmon farming in Scotland. On the surface the idea of farming salmon, perhaps to reduce demand for the rapidly diminishing wild stocks, seems a good one, but beneath the surface it's a grim apocalypse fuelled by power and profits.

The wild and beautiful silvery Atlantic salmon is a UK priority species, which is swimming rapidly towards endangered status in our lifetimes; it's thought that there are two-thirds fewer salmon entering UK rivers than there were just 30 years ago. While up to 35 per cent of young salmon entering the sea used to return to our rivers to breed, on most rivers now that figure could be under 5 per cent. Elsewhere the fish is widely distributed across Europe, ranging from Portugal in the south to Sweden and Finland in the north, but despite those declines the UK's population still comprises a significant proportion of the total European stock, and Scotland's rivers are a particularly important European stronghold for the species. And the only reason that Atlantic salmon persist here in any number at all is that a quarter of a million salmon used to be netted around our coastline as they returned to spawn, but now many of these netting operations have been shut down on conservation grounds. This must have helped, but it hasn't fixed the problem as the extra spawning salmon have simply vanished. Let's be clear though, those fishermen and women on the banks are not in any way to blame; they have been at the forefront of the conservation efforts and many now voluntarily practise tag and release – their fishies are not so often heading for the dishies.

The River Test runs through Romsey, a small town a few miles from my home. On an appropriately rainy Sunday in the mid-seventies I cycled there on an adventure to see wild salmon jumping for the first time. I'd seen it on TV and some grubby black and white photos in the local press. I arrived drenched, and the place was deserted except for one old man in a black mac. I was wholly socially inept as a teenager, so I parked my dripping bike and leaned over the railing around the churning pool some distance away from him. But it was instantly obvious that he had the prime spot, so I edged closer each time one of the fish rose stoically from the roiling spume. Finally we were just a few metres apart and I was just a few metres from the weir. The deafening roar of the waters, the anticipation of the next fish being bigger and closer, the exhilaration of the glimpse of its steely sides as it thrust itself into the wall of water remain fresh in my mind, but they are not the most vivid memory of that first sodden salmon watch. Sensing my presence, the man turned, smiled and beckoned me to take his place in the front row of this great natural spectacle. I grunted 'thank you', I was polite if not at ease, and immersed myself in the moment. After a couple of 'tiddlers' had bounced off the torrent, a monstrous dark fish soared up the cataract so close that I saw its eye filled with madness and desperation. I turned to return the smile to the man, but he had gone. He had given the biggest catch to me. He'll be long, long gone now, but there's no doubt his gesture is part of why I'm telling you about *Salmo salar*.

The salmon is a remarkable animal. It's anadromous, meaning that adults migrate from the sea to breed in fresh water. This necessitates phenomenal physiological changes, the most critical of which is the ability to osmoregulate. This involves maintaining a constant osmotic pressure in the body fluids through the active control of water and salt concentrations in both freshwater and marine environments, and being able to do so at precisely prescribed times. Why bother? Well, the river is full of predators and not that much food to feed a small fish that needs to be a big fish in order to produce enough eggs to produce enough small fish to guarantee there are more big fish. Hence they move to the sea, where they can scoff a lot and not get scoffed themselves. How did this strategy evolve? Not now, that's another volume of conjecture! When the salmon return they spawn in shallow gravelly areas in clean and fast-flowing rivers and streams where the water is full of life-giving oxygen. The eggs hatch and the tiny fish disperse into the river and, after a period of between one and six years, depending on their food supply, the young salmon travel downstream to the sea as 'smolts'. Then, after up to three years of feasting in the sea, they have an astonishing homing instinct that pulls them back to spawn in the very river of their birth. This behaviour has resulted in genetical differences in fish between rivers, even within individual rivers, and in fact even within the tributaries of single large rivers. Quite an animal. And plenty of people eat them too, as when properly prepared they can be tasty and are perceived to be a healthy food. But as they are becoming rare the answer has been to satisfy demand by farming them in captivity.

Salmon farming originated in Europe in the second half of the eighteenth century, and hatcheries were established a century later and on a significant scale in the 1950s in Japan, the USSR, the United States and Canada. The modern technique of farming salmon in floating sea cages was initiated in Norway in the late 1960s, and by the mid-1990s salmon aquaculture had overtaken the salmon fishing industry as the most important supplier of salmon worldwide. By 2004 the global production of farmed salmon exceeded wild harvests by more than one million metric tonnes. That's a lot of fish and a lot of money, and when there's lots of money to be made people get greedy, cut corners, intensify to increase their profits and can forget about animal welfare and the environment.

The Scottish Salmon Producers Organisation is a trade body that claims to represent every company farming salmon in Scotland and to 'help create the conditions for the long-term, sustainable growth of the sector and to give more consumers, at home and abroad, the chance to enjoy this world renowned, low-fat, healthy protein'. In April 2019 they published a report that claimed that the industry's turnover had been just over one billion pounds in 2018, and that it had also contributed £76 million in wages. International exports of Scottish salmon were said to be over half a billion pounds each year – which would equate to over half the value of the country's food and drink exports – excluding Scotch whisky. This in turn suggests that the economic impact of the industry might represent nearly £2 billion of turnover in Scotland's economy and, critically, that the total tax revenues generated would be £216 million. It's worth noting that salmon production is undertaken by around only twelve businesses who farm on about 225 active sites and that they employ only around 1,500 people. That's a lot of money in the pockets of not many companies and not many people. And a lot of tax in the government coffers. This is big business.

In this case big business relies upon big numbers of fish, and hundreds of thousands are raised in pens suspended in the open sea lochs stretching down the west coast of Scotland, as well as on Orkney and Shetland. The salmon swim around inside the pens for up to two years before being caught up for our canapes, sandwiches and dinners. Obviously they need feeding and they are sustained on a diet of processed food and constantly treated with medicines to ward off disease and infestations such as sea lice, which can proliferate among the high densities of fish in the cages. The pens stop the fish from getting out, but they allow these parasites to get in. They also cannot prevent thousands of tonnes of feed waste and faeces escaping into the surrounding water. The industry is hyper-intensive, but being offshore in remote areas it is largely out-of-sight and out-of-minds of consumers. Yes, the magnificent rugged west coast of Scotland, famed for its spectacular land-scapes, and considered a last vestige of pristine wilderness, is only a holiday destination for most, and few duck beneath the waves. Standing on those windswept shores most visitors, and perhaps you as a consumer, are unaware that every year millions of salmon die in those farms, reaching a peak in 2019 of about ten million fish. Diseases, parasites and even the medicines applied to treat them can all prove fatal. Sea lice have been seen to occur at such high densities that they are actually just eating the fish alive. Scared, skinless and grotesquely disfigured by raw lesions; you wouldn't want these poor creatures on your plate. Think about intensive battery chicken farming, only with fish and a direct line with our marine environment. Imagine if this was on land; imagine if our cattle were covered in parasites and dropping dead in the fields – it would never be tolerated.

So that is an ugly and unacceptable welfare issue, but is there any impact on those declining populations of wild fish and the wider environment? Marine biologists have discovered excessive parasite loads on the remaining wild salmon when they have passed close to the vicinity of the salmon farming pens. In the summer of 2018 Paul Hopper, a biologist with the Outer Hebrides Fisheries Trust, was exploring the Blackwater River on Lewis when he found a dead wild salmon with 747 lice attached to its gills, head and back. Marine Scotland, the government agency responsible for assessing the welfare of farmed fish, visited a farm nearby and rated it 'satisfactory' for parasites. Shortly afterwards a diver filmed the fish in the pens and recorded severe sea lice infestations. He published his images. Marine Scotland returned and found high levels of these parasites, which they described as 'nothing out of the ordinary'. Proprietors were sent warning letters. Other biologists have been looking at and recording the seabed beneath the cages. That's not a pretty sight either; a thick black and grey lifeless sludge, a bacterial mat formed when the feed, faeces and chemicals fall through the nets and kill all the marine life underneath. So who is meant to regulate this? It's the job of the Scottish Environment Protection Agency (SEPA) but, like Marine Scotland, they have been roundly criticised for their lack of action. Their own data suggest that fish farms are the second worst of Scotland's polluting industries and when the BBC's *Panorama* investigated Scottish salmon farming in 2019, 56 farms were rated by SEPA as 'very poor', 'poor' or 'at risk' by their own standards. Yet in the previous five years no company had been fined or prosecuted. A spokesperson for SEPA said that there had been 44 times in that period when they had directed companies to reduce the number of fish they had in their farms, but portentously added that this can cost them hundreds of thousands or potentially millions of pounds.

There are also concerns around human health when so-called 'pharmed' salmon is eaten. Alarm bells were raised in the US when in 2003 a report from the Environmental Working Group found that seven out of ten farmed salmon purchased in grocery stores in San Francisco, Washington DC and Portland were contaminated with cancer-causing polychlorinated biphenyls (PCBs) at 'levels that raise health concerns.' A year later a scientific paper in the prestigious journal *Science,* 'Global Assessment of Organic Contaminants in Farmed Salmon', reported that Scottish farmed salmon was the most contaminated in terms of cancer-causing chemicals such as PCBs, dioxins, DDT and dieldrin. Also without the addition of artificial colouring such as Canthaxanthin, 'pharmed' salmon would be a greasy grey in colour – not so appealing – so it's essentially given a fake tan. Yet this chemical has been linked to eye defects and its use was drastically reduced in 2003 following a European intervention on medical grounds. And, finally, SEPA got themselves into more trouble when they allegedly suppressed a damning 2016 report about the use of another chemical used in salmon 'pharming', emamectin benzoate, which is used to treat sea lice, but has lethal impacts on shellfish including lobsters, reducing their diversity and abundance by 63 per cent and 96 per cent respectively. FOI requests and the work of investigative journalists from *The Ferret* revealed that the report was buried after lobbying from the salmon farming industry, which argued that any publicity 'could undermine commercial confidence in the industry' and 'damage all of our reputations'. In January 2017 *The Sunday Times* revealed that the use of toxic chemicals on Scottish salmon farms had increased by over 1,000 per cent between 2005 and 2015. Never mind the sea life then.

Megs went to Scotland in 2018 to see some seals. They were lying on beaches adjacent to salmon farming operations, not resting and recuperating after a hard day's fishing – no, they all had bullet holes in their heads. The problem being that everyone loves a free inexhaustible lunch and seals are no exception. They are inexorably drawn to the salmon pens, where they can tear open the nets and help themselves to a feast. Until they get a bullet between the eyes. The latest data shows that 31 seals were shot under licence in the first three months of 2020, double that of the previous year. Many more are shot illegally. In other parts of the world they represent no threat as the salmon pens are mandatorily wrapped in protective steel mesh, but not so in Scotland – apparently the billion-pound industry wants to avoid the capital expense of fitting them. What they do use are acoustic deterrent devices (ADDs) that emit sounds underwater to scare off seals, this despite Scottish Natural Heritage, the government's conservation agency, clearly stating that they can cause hearing damage and stress in dolphins, porpoises and whales, and therefore are in breach of legislation to protect cetaceans.

Thankfully in June of this year, under the same legislation outlined above to one day protect mountain hares, the government announced that salmon farmers in Scotland will be banned from shooting seals and face strict controls in regard to the use of ADDs. Was it the animal welfare and conservation issues that prompted this – albeit belated – action? No, it was down to fears that the US could ban Scottish salmon imports because killing and harming seals will breach US regulations protecting the welfare of marine mammals in the wild. You see, America imports 25,000 tonnes of Scottish salmon; that's about 26 per cent of Scotland's export market, worth £179 million per year. Money talks. In fact, in regard to this industry, it's shouting very loudly. Everything written above is in the public domain and there are many exposés regarding animal welfare, human health, environmental damage, ecological impacts, conflicts of interest, etc, so it might strike you as frankly astonishing that the Scottish government wants to increase the number of salmon produced. Yes, unbelievably, they have set a target to expand production to 210,000 tonnes by 2020, and 300,000 by 2030. Despite increasing opposition from local communities, new fish farms are constantly being proposed and farms already in operation are seeking permission to increase in size. Maybe it's that £2 billion of turnover in Scotland's economy and those £216 million tax revenues ... who knows, eh?

WAR ON WASPS

Each summer there are many people who enter into a wildlife war of their very own. It's a war that lasts the season, but the battles can be numerous. The 'attacks' from the 'enemy' always seem to come when you least expect it – while enjoying a Mr Whippy or an ice-cold pint in the sun. You hear the buzz and you know it's on. Some get poised, ready to start their defensive swatting, while others dramatically retreat shrieking as they run for their ... for their ... um ... well, for what I am not quite sure. So, let's talk about the insects that strike terror into our hearts.

Wasps. In a study across 46 countries looking at human tolerance and perception of insects, wasps consistently came in last, with words like 'annoying' and 'dangerous' used to describe them. The root of our frustrations towards them stems from their interest in sharing a nibble of our sugary foods and, of course, the sting they give us when aggravated or threatened. At the opposite end of the scale are the bees, which we simply adore – even though they can give a pretty good sting themselves. Bees have become a symbol of a healthy environment, as they go about pollinating over 80 per cent of the world's plants, including over 90 species of crops that we depend on so heavily. But what purpose do the wasps have?

Well, wasps are actually just as ecologically important as bees. Wasps are both generalist and specialist pollinators, transferring pollen from flower to flower, supporting global food security. For example, fig trees and specialist fig wasps have co-evolved together over 60 million years and depend on one another to complete their lifecycles. Fig wasps pollinate figs (as you might have guessed), which are a keystone plant species that supports over 1,274 different mammal and bird species. But other than pollinating, they are also the masters of insect population regulation. Kathrin Klinkusch at the German Nature Conservation Union says, 'wasps are a health police; they eat other sick animals and maintain a healthy balance in the ecosystem.' Wasps control populations by preying upon vast quantities of insects that would otherwise overrun the countryside. Social wasps are estimated to consume up to 14 million kilograms of insect biomass, like spiders, greenflies and mosquitoes, each year. This helps to protect crops but also to reduce disease transmission. So we have got to re-think our attitudes when it comes to these creatures, which are seriously misunderstood. Without them, the world would be a much bleaker place.

You may not have been aware of the ills of the ill salmon gyrating in those toxic overcrowded cages, or of those hideous piles of decaying mountain hares dumped in our uplands, but I'm sure you'll know something about the badger cull. That's why I'm not going to detail its rationale (flawed), its practices (inhumane) and its efficacy (negligible to none) here. I'll point you instead to Dominic Dyer's excellent 2016 treatise on the cull, *Badgered to Death: The People and Politics of the Badger Cull*. It has all you need to know from a man who has fought ferociously and resolutely to end this vile anomaly. In the time since its publication the news only gets worse in that the cull zones have been extended, have spread to new parts of the UK and thus thousands more badgers have been killed, many in an inhumane fashion. Indeed in 2019 the government approved culling in eleven new areas, taking the total up to 43 zones that now cover 90 per cent of Wiltshire, 84 per cent of Devon, and 83 per cent of Cornwall. Their data now suggests that in the winter/spring of 2019 to 2020 a total of 35,034 badgers were killed and that by the end of 2020 the figure could reach 64,000. Dominic said, 'This year will take the number of badgers killed since the cull started to over 130,000, pushing the species to the verge of local extinction in areas of England that it has inhabited since the ice age', and with his typical pertinent clarity adds that, 'the public costs of the badger cull are estimated to exceed £60 million by the end of 2019, yet the government has provided no evidence to prove this cruel slaughter is having any significant impact on lowering bovine TB.'

Dominic is an ardent campaigner, so what about the scientists? Well, Professor Rosie Woodroffe at the Zoological Society of London, who has been working on badgers and the culling trials for years, said, 'I cannot understand why the government has permitted this massive expansion of badger culling, when it has not yet responded to the review it commissioned and received nearly a year ago'. That review by Professor Sir Charles Godfray, published in 2018, concluded that the government and farming industry were paying far too much attention to badger management, and far too little attention to cattle-to-cattle transmission, which is responsible for the majority of TB incidents in those cattle. Notably the review also called upon government to properly evaluate badger vaccination as a non-lethal alternative to culling.

The government did finally respond to Godfray's recommendations in March 2020, and positively, indicating that the cull would begin to be phased out in the next few years and that vaccination of badgers would be increased instead. Dyer was happy: 'Today the government has finally come up with a long-term exit strategy from badger culling based on cattle-based control measures and TB vaccination in both badgers and cattle. This is better for taxpayers, farmers and the future of our precious wildlife.' Rosie was happy: 'This is a seismic shift in an area of government policy which has been highly controversial for many years', she said, adding, 'The response rightly highlights cattle-based measures and this focus is appropriate, because the best estimates show that most cattle herds that acquire TB are infected by other cattle herds.' And the rest of us in the UK who care about wildlife and its welfare were happy too. That was March; fast forward to May ... when the same government approved seven new badger culling sites in new areas that cover parts of Gloucestershire, Herefordshire, Devon, Dorset and Cornwall. So the killing goes on.

The badger cull will always be a revolting stain on the UK's reputation as a world leader in the field of wildlife conservation. Many have tried valiantly to stop it, like Dominic and Rosie, and Brian May and his Save Me Trust team, like all the cull monitors who have gone out in the cold, dark and wet, all those who have marched, fundraised and signed petitions. Sadly, other ostensibly prominent players who purport to work to protect our wildlife haven't tried as hard as they should have. Whatever; we lost, and thousands of badgers have died at the behest of the ignorant, greedy and stupid, something we will all have to live with for the rest of our lives.

Once again we have to see ourselves as the last bastion of defence; that's you and me. We can complain and campaign, we can continue to resist this slaughter by joining those on the front line, at night in the cold and dark and wet to monitor the cull, we can sign the petitions and give whatever we can afford to support badger vaccination programmes, we can celebrate these remarkable animals to raise awareness of the unjust and unscientific nature of this abuse and, if we are lucky, we can nurture and protect those badgers that share our communities. Go out at night, sneak into their world, hold your breath, peer into their shadows, listen to their snuffling and scratching and feel your heart miss a beat when one passes close to you. When you share intimacy with a creature you feel an affinity for a creature, and when affinity grows it becomes a part of you, a part you don't want culled.

In conclusion, wildlife in the UK is under severe threat and what we've seen above is that our system of government continues to allow the power to remain resolutely in the control of the shooting, fishing and farming industries, some of which are doing severe damage. Sometimes it feels that wherever we look we can see the catastrophic impact of putting the interests of these sectors above protecting that wildlife. Money, economics, jobs, livelihoods – all are frequently touted as justification to keep things the same, to keep business as usual, but as we've seen, there are workable alternatives. Tradition is another excuse, but for me traditions must be robustly questioned if they don't have contemporary relevance. Look at the abominable spectre of fox-hunting, surely the heir apparent to football hooliganism as the 'English Disease'. That being said, despite being one of the most popular pieces of legislation on the statute books the

Hunting Act of 2004 remains a target for the hunting lobby, who insidiously attempt to use their political influence to scrap the law – which is constantly abused anyway. They claim that 'trail hunting' is a legal activity, but very often it's just a shabby ruse to mask the active hunting of foxes and, unbelievably and shamefully, the illegal hunting of wild animals with dogs remains common across our countryside. It's appalling.

NO P-E-S-T-S HERE, ONLY FRIENDS

Most of us love them, some of us even feed them, a few of us are indifferent and a handful will chase them with hounds over the countryside enjoying a good old 'traditional' day out as they run terrified for their lives. I will never understand or sympathise with that way of thinking, especially since we now have so much information that tells of foxes' ecological importance and their alluring nature.

Red foxes are one of the wildlife gems of the UK. They have adapted to live in cities, and in those environments they are increasingly generalist in their behaviour. An urban fox's diet consists of 55 per cent natural prey, like small birds and rodents, 20 per cent insects and earthworms, 7 per cent fruit, and the remaining 18 per cent is made up from scavenging the waste we leave behind. In the 1990s, only 33,000 foxes were estimated to live in our cities, whereas today there are in excess of 150,000. Due to their close proximity, foxes have been very widely studied. One infamous investigation in Russia in the 1960s attempted to domesticate foxes by breeding the least aggressive individuals each generation for ten years. Towards the end, the foxes developed shorter snouts, floppy ears and even a dog-like bark. This was, of course, an artificial study, but now something very similar seems to be happening naturally within the northern cities of the UK.

New research suggests that foxes are evolving dog-like traits that are potentially leading to some form of self-domestication. Kevin Parsons, an evolutionary biologist at the University of Glasgow, was taken by the number of interactions he had on the streets with foxes that would walk by him and simply stop and watch. Parsons began comparing fox skulls and found that urban foxes were evolving wider and shorter snouts with smaller brains than their rural counterparts, an effect termed by Charles Darwin domestication syndrome. It's hypothesised that urban foxes are developing a skull designed to have a stronger bite to better crush the bones and hard foods left out by people. Foxes are far from being truly domesticated, but it is interesting to see how they are evolving to live alongside us. These animals deserve our respect and compassion. They're resourceful, have strong social bonds and are anything but ... p***s! P*e*s*t*s!

So what can we do? Firstly, vote in a government that manifests a real concert for wildlife welfare, but in the meantime we should instantly ban the use of dogs below ground by hunts, which leads to the death of foxes, and we should add a 'reckless provision' clause into the Hunting Act to stop hunters hiding behind the pretence of trail hunting. We should put an end to the unnecessary and time-consuming consultations with so-called stakeholders, which are merely there to facilitate their nefarious objectives and water down what really needs fixing. And we should put a stop to the enduring delays once decisions that benefit wildlife have been made – if we are going to stop shooting seals and mountain hares just ... stop it!

When it comes to wildlife crime, we should make all incidents recordable offences because a clear consequence of under-recording is an inconsistent approach to police investigations. If the figures aren't there to prove that wildlife crime is a problem, then the police will struggle to devote funding and resources from their already over-stretched budgets to properly address this issue. That's why it should be compulsory for the English and Welsh governments to publish an annual wildlife crime report, as it is in Scotland. Policing wildlife crime also needs to be proactive rather than reactive. In the fight against drugs we don't sit back and wait for a crime. Officers should be out in known wildlife crime hotspots conducting undercover investigations. And who are the real criminals? Are they the employees or the employers setting the agenda for their illegal activities? Corporate liability in the form of vicarious liability should be an offence in England and Wales as it is in Scotland. And remember the Stody Estate? Estates where employees commit wildlife crimes should automatically lose *all* their subsidies for landholdings. Lastly, we need to shake up the judiciary and let them know that these crimes are serious and therefore warrant a substantial increase in penalties, and perhaps additional penalties for crimes with conservation impact or those committed inside national parks – remember that poisoned eagle in the Cairngorms? If you get caught, then bang, there go your firearms and shotgun certificates for a minimum of ten years. When it comes to dealing with wildlife conflicts in the UK we need to wake up and get serious. We are already mad, so now is the time to get even.

Langholm Moor Buyout: Communities in action

Since the nineteenth century, Langholm Moor, owned by Buccleuch Estate, has been managed primarily for grouse shooting. The area covers 25,000 acres (the same size as 12,500 football pitches), positioned between Dumfries, Galloway and the Scottish Borders, and is famed for its 25-year study investigating the survival of hen harriers on grouse moors, the results of which indicated that hen harrier numbers increased when the moors were managed by gamekeepers for grouse. But there are two big problems here – firstly, it is not good practice to build an artificial population that is sustained almost solely by the presence of red grouse, which harriers will feed upon, and, secondly, because all other predators that might take chicks – primarily foxes – had been controlled (i.e. legally killed). When this unbalanced moor was no longer intensively managed, the numbers plummeted as the population was unable to sustain itself naturally due to the impaired landscape and lack of food. But there is a solution to this that could promote a sustainable, healthy and natural hen harrier population.

In May 2019, Buccleuch Estate announced it was planning on selling Langholm Moor. This was partly because the land was no longer viable to hold a premier grouse moor as it lost half its heather to sheep grazing and became constricted by conifer plantations that fragmented the area and limited the success of red grouse, which depend on young heather. The Duke of Buccleuch was once one of the UK's largest landowners and was given his hereditary title that dates all the way back to 1663. The family holds a total of 217,000 acres of land and lots of urban properties, but for the first time in generations they are selling part of their estate.

This was an unique opportunity that was noted by the residents of Langholm, who have devised a modern and ambitious plan to purchase 10,500 acres of the moor as a community to reform and restore it to its former glory, before humans decided to poke their noses (and guns) in. Instead of a managed mono-culture of heather that contributes to the climate crisis, imagine a scene of ancient woodland, peatland and grasslands supporting a balanced ecosystem of raptors, songbirds, mammals, reptiles and insects. You're standing there watching as hen harriers dance in high numbers, made possible by the availability of natural food sources. You can hear hedgehogs scuttling in the undergrowth and foxes calling to one another as the sun sets on a wildlife utopia. It's a beautiful, safe place for ethical tourism and learning. Well, that's the hope of the Langholm community. They aim to turn an old, drab, degraded grouse moor into something fresh and exciting – the Tarra Valley Nature Reserve. By engaging in ecological restoration, wildlife conservation, climate solutions and community regeneration, the overarching goal is to improve the area for both the environment and the local economy. The land would be owned by a local charity, the Langholm Initiative, which is governed by a voluntary board of community members. Much of the land is already a Site of Special Scientific Interest and a Special Protected Area. In the first five years, the community hopes that the new reserve will be awarded National Nature Reserve status.

This idea didn't just evolve from a whim. Project leader Kevin Cumming said, 'We aim to demonstrate that nature and community regeneration can go hand in hand. Hopefully this project can act as a blueprint for others to follow. We are at a critical stage in tackling the climate crisis and it has never been more important to demonstrate to our children that we must do better, we can do better and we will do better, for their sake.' The Langholm community raised funding to thoroughly research the moors' potential and the results showed that the creation of a reserve would be hugely beneficial to both wildlife and people. There is one pretty substantial hurdle to overcome to make this project a reality – yeah, you guessed it – it's money ... a whole lot of money. But more on this in a bit.

One of the communities' main intentions is to initiate a large-scale peatland restoration project to benefit biodiversity and to capture carbon from the atmosphere. Around the world there are 46 million hectares of degraded peatland, meaning that through draining and burning (as is done on grouse moors) the carbon that was once stored in plant material within peat soil is released. Global restoration schemes would prevent 394 million tonnes of CO_2 *entering the environment each year – the equivalent to 84 million passenger vehicles. In England, peatlands cover 11 per cent of the land area and hold an estimated 584 million tonnes of carbon within their soil. This habitat also plays an important role in maintaining water quality and supporting wildlife. However, in the last 300 years or so our peatlands have been significantly damaged and destroyed due to the actions of humans. Natural England studied the extent of this and found that, shockingly, less than 1 per cent of our peatland habitat remains entirely undamaged. This is due to irresponsible intensive agriculture, unnecessary burning for grouse moors, pollution, draining and direct peat extraction. What an absolute mess. A total disgrace. Large-scale and long-term restoration projects, like that proposed by the community of Langholm, could reduce emissions by up to 2.4 million tonnes of* CO_2 *annually, helping to mitigate against the climate crisis. The community hopes to block the numerous peatland drains on site to allow for the return of raised bog that will promote biodiversity, prevent flooding of local areas and remove additional carbon. In addition, they plan to plant 500 acres of native woodland designed with open glades and wet zones that is easily accessible to visitors and promotes the ecological integrity of the landscape. Count us in!*

Langholm was known as one of the UK's most productive and important textile manufacturers, specifically spinning woollen yarn to create one of Scotland's most iconic looks – tweed. With the rise of what we now call fast fashion, the pressure on these mills to keep up was astronomical and the mill announced its closure as a result of the Covid-19 pandemic in 2020. Like other similar small communities, Langholm suffered a major reduction in employment opportunities, as well as the pressures of an aging population, lack of diverse industry and little investment from housing or business developments. If the community is able to purchase Langholm Moor and reform it into the Tarras Valley Nature Reserve, it would have huge human benefits too. There could be a field studies centre for students to build upon their environmental knowledge and skill sets, opportunities for responsible tourism and many more job opportunities for local people. Within their proposal it says, 'the community has never before had a say in its own future, whether that is the development of industry or how the surrounding landscape could be used to benefit the town itself ... until now.'

To me, this sounds like a project full of heart, which is backed up by sound science that has the potential to right some of the wrongs we have caused to our environment. When given the space and time to reform, nature will come back and prosper. There is just one tiny little hiccup to this well-constructed and brilliant proposal, and that little hiccup is that Buccleuch Estate are selling their land for around six million pounds. Land in the UK is tremendously expensive and is always one of the main issues when it comes to conservation. The community group, led by project leader Kevin Cumming, needs at least £3.5 million from crowdfunding or private donations, as well as seeking the financial support of the Scottish Land Fund. It is ambitious, but Rome wasn't built in a day. This is an opportunity to take back the land and regrow a healthy environment that was so grossly stolen away through consistent burning, draining, shooting and trapping. It should have been there for all to enjoy and to learn from all along, yet it has been missing for so long that most of us don't even recognise what's been lost. Finally, in the wise words of Alan Watson Featherstone, founder of Trees for Life, 'ecological restoration, as the human-assisted recovery of degraded ecosystems, is in essence the work of hope.' We are at a tipping point, but we don't have to fall over that edge if we don't want to.

WHO'S AFRAID OF BIG BAD CONSERVATION?

The big bad wolf slips silently down the precinct paving and into the alley between the supermarket and a row of shuttered shops. It's hunting; it stops to sniff the smelly city, straining for a whiff of its favourite prey, the lavender pot pourris, the tart perfume of blue rinse and the other stereotypical scents of ... grandmothers. Of course, it could get luckier still, and fall upon that other staple of its diet – a child in a red caped coat.

What is it with people and wolves? Why is it that ecologists can easily educate people about the irrefutable value of the honeybee, get people to swoon over puffins and dormice, but never get them past the fairytale image of this remarkable and essential predator? It's infuriating and irrational, and that irrationality sends me scrabbling for some sort of reason. So personally, I think it's down to genetics. Humans have co-evolved with dogs, every last one a descendant of the wolf, to the extent that we are born with a fundamental understanding of this creature, and unsurprisingly, them of us too. And given that they were domesticated between 35,000 and 17,000 years ago, and that long before and throughout we have been living in fear of their progenitors, maybe our irrational mistrust in *Canis lupus* is rooted in our DNA.

I'm afraid however that in the twenty-first century this provides no adequate excuse to any tabloid copy writer, land-owner, hunter or farmer. We know better. We know that keystone predators are not only desirable but essential when it comes to sustaining healthy ecosystems. Surely we can over-come our primal insecurities and be reasonable? Well, no, evidently, we can't, and for that reason linking the reintroduc-tion of wolves to the case for rewilding has been a PR disaster. For many the opening fanfare heralding the return of large predators was too easily scary, ludicrous, irresponsible and the perfect ammunition to fabricate a case against this exciting and increasingly necessary conservation process. We aimed too high, trusted too much and underestimated the ignorance and animosity of our adversaries. For these reprobates, wolves are evil, lynx are feline tyrannosaurs and, worse, beavers are a 'pest' before they've even got started. So conservationists need to learn to be more cunning, wily, sly. We should plot in packs, not pander to the plonkers.

In 2013 George Monbiot's much-lauded or loathed book *Feral* was published, bringing the whole concept of rewild-ing to wider public attention, but in truth rewilding had been happening before then, albeit on smaller scales. The RSPB had taken a carrot field and turned it into Lakenheath Fen in quick time, rapidly making a home for bitterns, bearded tits, marsh harriers and common cranes. But while this was more of an experiment, Monbiot's lyrical polemic espoused a vision, one he typically validated with rigorous research, the right science and irrefutable statistics. Monbiot didn't make a nature reserve, he lit a torch and held up an environmental manifesto that offered constructive solutions to ecological problems and offered to rejuvenate our ruined landscapes. 'We live in a shadowland,' he wrote, 'a dim, flattened relic of what there once was, of what there could be again'. To me it was exhilarating and should have become an explosive catalyst for a brave new environmental movement. But his exquisite vision demanded more courage than the UK's conservation movement could muster and his term 'sheep-wrecked' – coined to very accurately describe the overgrazed and over-subsidised wastes of upland Britain – terrified the more inflexible and intransigent factions of the farming fraternity. The former hid under its covers and the latter blanketed his beautiful idealism in scorn and hatred. Ho hum.

The charity Rewilding Britain is good. They promote the idea and sell the dream, but here in the UK we have a very fundamental problem with how to effect proper large-scale conservation: land. More specifically, who owns it and how much it costs. Guy Shrubsole's book *Who Owns England?* lays bare our horrid predicament. Of course it's a lovely rant about history and politics; thus it recounts how, when asked how young entrepreneurs might succeed in the UK, the late Duke of Westminster, then-owner of a considerable slice of London's Belgravia and central Liverpool, uttered pithily that they should 'make sure they have an ancestor who was a very close friend of William the Conqueror'. Because it hasn't really changed since 1066. Because we haven't had the land reforms we should have had and, as a result, 40 million acres – that's two-thirds of the land in the UK as a whole – is owned by 0.36 per cent of the population. The rest of us, that's 24 million families, are squeezed into the 'urban plot' of just three million acres. And, as an aside, did you know that ten times more land is given over to golf courses than to allotments? And so when it comes to nature reserves we are effectively dealing with 'allotments', and we simply can't afford to buy enough to make a difference. In 2014 a plot of boggy mud on Dorset's Sandbanks peninsula, which measured 27 by 18 metres, was offered for £6.25 million. There's not enough room there for a sand lizard, a family of ladybird spiders or half a smooth snake – not that the neighbours would want them anyway. More realistically, prime arable land in 2019 was going for around £9,000 per acre, but individual plots could easily reach £15,000 per acre. Ask yourself what species could live securely on just one acre of land – that's one football pitch or sixteen tennis courts. Don't strain, I'll tell you: one or two species of sedentary mollusc – until a fire wipes them out, or the government requisitions their fragile patch to build a high-speed railway line.

Now, go onto the Rewilding Europe website and swoon. Feel the envy, admire the progress, enjoy the success. It's a beautiful thing to make beauty, to be bold and brave and brilliant. They have their own bank, for Christ's sake! Rewilding Europe Capital loans money to ecological entrepreneurs very cheaply to encourage sustainable development in their eight fabulous action zones, from Swedish Lapland to the Danube delta. They have a 'Bison bank' too – yes, a herd of full-on primal grazing animals that you can borrow from to get your rewilded ecosystem going. I so wish I had enough land to borrow a bison. But let's get back to the money ...

The bank is Europe's first rewilding enterprise fund that provides financial loans to 'any commercial activity or business that generates economic or social benefits in ways that support and achieve the generation of both rewilding outcomes and natural rewilding impacts'. Okay, but how much? Well, they've raised €7.5 million of principal funding and, at the time of writing, have made 21 loans to the tune of €2,020,000 to 152 enterprises, all of which are currently performing positively. Why is this not happening in the UK? Because sadly there is currently no consciousness of the commercial potential of actively engaging with and supporting rewilding. Its adversaries will tell you that it will mean a loss of jobs, that forests and marshes won't pay more than sheep, and so through their mischief in these rural areas the commercial value of natural capital is very poorly understood and access to development finance elsewhere is virtually non-existent, meaning nature-based exercises remain thin on the barren ground.

In November 2019 *Farmers Weekly* reported that many upland sheep farms in Wales would not survive without the subsidies they receive from the taxpayer. The study, conducted by Aberystwyth University, revealed that the payments, together with diversified income, contributed to about 45 per cent of turnover and 242 per cent of profits for these hill farms. After rent and finance charges were deducted, the survey showed that the average profit for hill sheep farms was £158 per hectare, with income from subsidies accounting for £243 per hectare. The study director remarked that, 'It is difficult to see how many Welsh farms can be profitable without relying on non-farming income and post-Brexit support payments.' But the truth is they are not profitable, they are subsidised, and we pay more in subsidies than they make in profit. Again in 2019, European Union subsidies amounting to around £350 million a year made up on average more than 80 per cent of farm incomes in Wales, and in May of that year Agriland reported that the Welsh government farm income forecasts showed that average farm income was expected to decrease by 15 per cent, to £29,500 compared to 2018. Most worrying, Wales' biggest sector, sheep farming, was expected to see a massive drop of 29 per cent to an average farm income of just £17,000. It's a sorry state of affairs all round. For the poor farmers, for the taxpayers and for the 'sheep-wrecked' environment and its paucity of wildlife. Clearly we need a solution and maybe the National Sheep Association should think a bit more progressively about embracing the 'wolf' of rewilding to help its troubled members.

Rob Stoneman of Rewilding Europe thinks so ... 'What happens to those hill farmers? Do we as a society just abandon them, do we let them fall to their fate?' he says. 'Sheep farming isn't economic, so what are the alternatives? Rewilding is one of those ... If you did that you would create a really exciting landscape. A lot more woodrows, scrub, a heck of a lot more wildlife, lots more birds, mammals, big grazers like deer and wild boar, and maybe wild cats.' Okay, so that's all good for wildlife, but what about the people? He goes on, 'That's going to attract people in. We know that tourism is part of the global economy that is continuing to grow, and of tourism the fastest part of the niche is landscape, nature, activity-based tourism.' And guess who has set up a European Safari Company? Yep, Rewilding Europe, of course. And is it a bit more carbon friendly than East Africa? Er, what do you think? And do people get a great thrill out of tracking wolves, lynx, bison and bears across Europe, camping under the stars and eating local foods? Er, what do you think? And would they enjoy doing the same in Wales? Er, what do you think?

A long-time mate of mine, Derek Gow, is a brilliant conservationist and former sheep farmer. He got fed up with the fact that 'farming today is virtually entirely reliant on subsidies, especially when you come to the upland areas – you're not farming sheep, you're farming subsidies'. But he's not bleating on about it; his 120 acres at Upcott Grange Farm in Devon is now being rewilded. It's not a walk in the park, he admits, but it's now open to the public and taking tour bookings. Derek was a farmer and he had the vision to realise that, Brexit or not, farming is going to have to change, and the farmers who embrace that will be the first to prosper. So what of our government's reaction to farming's social and environmental tragedy? Here's what a DEFRA spokesperson had to say: 'When we leave the EU we will maintain the same funding for farm support until the end of this parliament and we have been clear we will intervene to provide additional support where necessary. We have been meeting regularly with the food and farming sectors across the UK for some time to understand and anticipate the potential impacts of a no-deal scenario on our agri-food industry. We are making all necessary preparations to ensure our farming industry is ready and that Brexit works for farmers across all parts of the UK.' What did I say about vision?

CHATTY TREES

Communication is key on so many different levels. In biological terms, communication encompasses sending out a signal, whether that be visual, auditory, olfactory (smell), chemical, hormonal or tactile touch. It's about the transfer of information, which is vital for the survival of most life-forms. These signals may deliver a warning or perhaps pass on knowledge of resources and skills. Animal communication has been fairly well studied over the years, although there is still so much left to learn. But plant communication, on the other hand, is fairly new and we are just scratching the surface of the magnitude of signals sent out beneath our feet under the soil.

Walking in the woods I am always struck by the beauty of the trees, standing tall and seemingly solitary from one another. But this couldn't be further from the truth as they are all connected by an underground fungi network. It is believed that all plants have some sort of symbiotic (mutually beneficial) relationship with fungus in the soil. These different fungi species connect tree roots together, creating a complex system known as the mycorrhizal network. The fungi benefits from consuming the glucose produced by trees during photosynthesis, and the fungi absorbs soil nutrients to give back to the trees. Scientists have nicknamed this the 'wood wide web' because the network also allows trees to communicate with each other, as the fungi network can transport nutrients from one tree to another. For example, young tree seedlings may be growing in the shade of a mature parental tree and therefore unable to gather enough energy from the sun to grow, so the big trees will share their nutrients in the fungal network. This can also happen across species. The Douglas fir and the paper birch trees are in leaf at different times of the year, so when one is leafless the other will transfer carbon and nitrogen as the tree in full foliage will have the best nutrient uptake. Maybe we need to take a leaf out of their book (pardon the pun – sorry, not sorry); collective communication and thinking provides so much more support and strength when it comes to conservation.

So where does rewilding in the UK go from here? Well, faced with the fact that we own no land, we certainly can't afford to buy enough to make a difference and we aren't going to get much help from the farmers' unions and the government, things seem pretty bleak. But they are not. Far from it in fact, because parts of West Sussex and the Scottish Highlands have been turned into an absolute utopia for people, for wildlife and for their economies. All thanks to visionaries, the first in the forms of Charlie Burrell and Isabella Tree, custodians of the Knepp Estate, south of Horsham.

If you google 'Knepp Estate' you get that panel with the Google reviews, the address, map, opening hours, etc. But you can also ask questions. A few months ago, someone typed in, 'I want to birdwatch in the grounds, I want to hear turtle doves'. The reply politely and pragmatically states, 'If you are looking to hear turtle doves then the Southern Block is the best area. You can park at the Walkers' Carpark on the route into New Barn Farm and the Rewilding Project. By exploring the footpaths and bridleways around the Southern Block and around Hammer Pond very early in the morning or late in the evening in the breeding season (June and July) you stand a good chance of hearing them.' I love that. I really love that, and do you know why? Because since 1970 the turtle dove population has declined by 95 per cent in the UK and now this 3,500-acre rewilded, formerly unprofitable farm is the only place in the UK where it is increasing. Ditto the nightingale, cuckoo and purple emperor butterfly. And we can go and listen to them – brilliant.

You really must read Isabella Tree's book, *Wilding*, which charts the history of the project and its early successes, but I'll summarise the gist here. Charlie inherits the farm in 1983 ... along with its debts, and for seventeen years tries to flog the land to pay his way out of them. He can't – it's not profitable. They have some beautiful old trees and seek advice about their conservation, which means they can't carry on ploughing their roots up ... so even less chance of a profit looms. In 2002 he has an epiphany and dreams of a process-led, non-goal-orientated project where, as far as possible, nature takes the driving seat. He tells Natural England he's going to establish a biodiverse wilderness area in the Low Weald of Sussex. They don't get it ... until 2010 when he finally gets some stewardship (subsidy) support. Now it represents a triumph of conservation, where issues such as soil restoration, flood mitigation, water and air purification, pollinating insects and carbon sequestration have all been addressed in parallel to an astonishing increase in biodiversity. And now, sixteen miles south of Gatwick Airport, sits the most beautiful part of England. No contest.

Charlie and Isabella's mentor is the Dutch grazing ecologist Frans Vera. Vera argues that we have forgotten that we once had huge herds of now-extinct grazing megafauna exacting a critical impact on our ecology. Not long ago, bison, elk, tarpan and aurochs (the original wild horses and cattle), as well as beaver, wild boar and red and roe deer, would have been munching away at the vegetation in vast numbers. Their browsing and grazing techniques, not to mention all the physical disturbance caused through puddling, rooting, debarking and the enormous redistribution of nutrients and seeds, would have produced a complex mosaic of habitats. The idea of a thick, old, primal closed-canopy forest smothering all of Europe is not one Vera subscribes to; he imagines a more open wood pasture, a shifting landscape of open-growing trees, scrub, grazing lawns and thorny thickets – all shaped by hungry herbivores. A kind of European savannah rich in biodiversity.

The ruminent herbivores had already been doing their thing in a pioneering experiment in the Netherlands on a reclaimed polder at Oostvaardersplassen – 56 square kilometres of fenced wetland into which Heck cattle and Konik ponies (substitutes for aurochs and tarpan) and red deer were introduced and left to their own devices. Unfortunately, it has proved highly controversial due to the ongoing number of animals starving to death when the populations grew too big for the area's resources. In the winter of 2017/2018 almost 3,300 deer, horses and cattle died, and this caused a justifiable commotion. Delve deeper than the headlines, though, and you'll find that politics and farmers prevented the establishment of corridors to other natural areas in the Netherlands and even Germany. It was a big, brave plan, but it ran out of money, and mismanagement and infighting caused it to fail. But there was a bigger and far simpler problem: all those herbivores, and no carnivores.

Without the obvious addition of wolves, lynx and brown bears there was an impact or, more specifically, a lack of it. It's not just about them killing and eating things, though; an 'ecology of fear' felt in the presence of predators radically changes the behaviour and grazing patterns of their prey. And ultimately it's the basic food resource, their forage, that will regulate their populations in terms of stress, disease and ultimately starvation. Culling would have been the obvious solution, but the managers and public couldn't agree on that either. Let's be clear, the project has not been a catastrophe in terms of experience, scientific research or the huge increases in biodiversity, but like the 'Rewilding = Wolves' headlines, it hasn't done this practice many favours in the eyes of the public.

So could Charlie do something similar, but avoid the pitfalls? Yes, because he made some pragmatic decisions, played to his strengths and was realistic. So, no large predators just south of Gatwick – that was a given – but his masterstroke was to use farm animals to replicate the work of the extinct or difficult-to-manage contemporary grazers. Swap the recalcitrant and dangerous Heck cattle for Old English Longhorns, the unruly Konik ponies for Exmoors, use Tamworth pigs instead of wild boar, bring in some red deer to join the fallow and roe. The key to generating and then maintaining a rich mosaic of habitats would very clearly be in managing the numbers of these animals. Too many and it would just be grassland, too few and it would be closed-canopy woodland. But remember, Charlie is a farmer, so he keeps the populations within these parameters by taking animals off the land to process into meat. Knepp sells 75 tonnes (live-weight) of free-roaming, pasture-fed organic meat every year – an important income stream for the estate. So, in reality, they are still farming, but on a very extensive scale and with very low carbon inputs – much like ranching. The dividends come in terms of biodiversity.

Gary Moore is a sound recordist and naturalist; you may have seen him eulogising about nightingales or tawny owls on *Springwatch*. He is not easily impressed, some would say a more glass-half-empty kind of bloke, though I'd say he's a realist. He visited Knepp this May and told me he'd been to heaven on Earth. Not surprising, as he'd recorded nightingale, turtle doves and cuckoos all at the same time. There's nowhere else in the UK he could do that in 2020.

The UK nightingale population has declined by 90 per cent since 1970. Once widespread across the UK, their rich and famous song inspired Shakespeare and Keats. But it's not just about its complexity and clarity: they're loud too, the males pumping up their volume to lure the migrating females down from the night skies. I haven't heard one this year – lockdown saw to that – but in the 1970s, while I was bopping to the Clash, they nested in the woods where I'm currently sat typing. Sadly, their range is now limited to the south-east corner of England and numbers continue to freefall.

Prior to the rewilding project a national nightingale survey by the British Trust for Ornithology recorded only nine nightingale territories at Knepp. But in 2012 the survey identified 34 territories and a year later that number had risen to 42 pairs in the Southern Block alone. It now has one of the highest densities of these birds in the UK. Remarkably, when it comes to the turtle dove Knepp is probably the only place this species is actually increasing. I haven't seen one in years. I've had the misery of watching them get blown out of the skies in Malta, where spring hunting of this sub-Saharan migrant continues despite declines of more than 96 per cent in the UK since 1970. And this sickening shooting of a species plummeting to extinction continues across Europe; 800,000 are killed in Spain each year. They will almost undoubtedly be the next bird to become extinct as a breeding species in the UK, and when they do the last gentle purring of the secretive male will be heard from a thicket on the Knepp estate. In Knepp's farming days there were none, by 1999 there were just three singing males, by 2019 there were 19. Catch them while you can.

But it's not just these two legendary songsters that have prospered here; populations of many common species have exploded and rarities have appeared too. Peregrines, ravens, red kites, lapwings, skylark, yellowhammer, stonechat and lesser spotted woodpeckers are all breeding, along with all five species of the UK's owls – and a few oddities pop in, such as nightjars, normally considered a heathland bird. Thirteen out of the 25 species of bat flit about, harvest mice and dormice weave their nests, hedgehogs have returned, and a total of more than 600 invertebrates have been recorded, including 37 out of our 59 butterflies.

INDIVIDUAL ACTIONS MATTER

We all have a drive to make our environment a better place, especially when it feels unsafe. However, we aren't the only species striving to make the world a better place – it's just that for some organisms their world may be very small by comparison to our own. But, to them, it is equally important and meaningful.

Rock ants are a species that you can find around the coasts of the UK and they have some very particular home specifications. They will only occupy crevices with low light levels, an entrance hole of about 1–1.5 millimetres, a ceiling height of approximately 2 millimetres and an internal living space of $20cm^2$. Not any crevice will do! But due to the coastal topography, colonies can be easily damaged by larger animals or rock slides. To minimise the risk, scout ants will always be searching the area for a better crevice to relocate the colony to. Thomas O'Shea-Wheller at the Ant Lab at the University of Bristol conducted experiments to better understand what it takes for a relocation to take place by presenting the scouts with good-, medium-and-poor quality crevices. O'Shea-Wheller explains, 'by remaining in a nest, ants effectively contribute to a quorum. The longer they spend there, the more likely it is other ants will join them.' However, he found that there was considerable variation between individual scouting ants as some were pickier than others. A few individuals never seem content with the crevice they are in – even if it is secure and fits the specifications to a tee they will continue searching for better. At the other end of the spectrum, some individuals would be accepting of any crevice, as long as it was an improvement on the original.

Consistent differences in behaviour can be attributed to personality, but it is a far stretch to say that individual ants have personality traits. Still, O'Shea-Wheller's results do show that there are individual differences when it comes to decision-making. Individuality is important to maintain the colony, as those fussy scout ants will often find and drive relocation to a better crevice, but ultimately it takes the whole colony to move and respond. There seems to be an ironic parallel here, don't you think?

Out of all of Knepp's successes perhaps the most astonishing is the purple emperor – from zero, to first seen in 2010, and now to the largest colony in the UK. And better still Knepp is not a great big thick oak woodland, the habitat we formerly thought this extraordinary insect required. What seems to have led to this super-colony is the post-rewilding abundance of sallow scrub, as it is this tree where the females lay their eggs and that the weird green caterpillars feast upon. And, best of all, you can go and see these ostentatious dandies each July by booking a safari or by staying in one of Knepp's tree houses, shepherd's huts, tents or yurts set in a wildflower meadow or on the edge of ancient woodland – you can pitch your own tent and glamping is part of their business model. It's not all cheap, but it's not the cost of a trip to the Serengeti and your carbon footprint would be minimal in comparison. Go there if you can, and be amazed.

Knepp is beautiful and a conservation triumph, a title it shares with the UK's other great rewilding initiative, Cairngorms Connect. This simply extraordinary project set in the Scottish Highlands is a fair match for those fabulous areas managed by

Rewilding Europe. Over 600 square kilometres of ancient Caledonian pine forests, wild rivers, lonely lochs, blanket and woodland bogs and the impressive mountain massif are home to eagles, ospreys, wildcats, pine martens, red squirrels, and rarities galore, from purple sandpipers to the silver stiletto fly, Kentish glory moth and shining guest ant, twinflower, intermediate wintergreen, lesser butterfly orchid, the list goes on ... It's a cornucopia of rare, enigmatic and beautiful species and habitats. For me, however, its greatest attraction is the fantastic partnership that shares the brave vision and its 200-year-old plan: the RSPB (an NGO charity), Forestry and Land Scotland – formerly Forestry Commission Scotland – and Scottish Natural Heritage (government agencies) and tellingly the Wildland Estate – a privately owned body of land belonging to the Danish philanthropist and clothing retailer Anders Holch Povlsen, a man whose hand I'd very much like to shake (metaphorically in these socially distanced times of course). Povlsen is the largest private landowner in the UK and has bought a large suite of estates across Scotland, reportedly totalling some 890 square kilometres (345 square miles), which he eventually plans to interconnect and rewild. Yes please!

Now you might find my enthusiasm contradictory, given the facts about land reform, or the lack of it, and land prices earlier in this chapter, but here is a highly motivated and clearly informed individual who can largely bypass the idiotic red tape that chokes so many conservation projects, and has the financial wherewithal to just get on with it – in 2019 he planted his four millionth tree! He's clearly not playing the benign dictator either as he has assembled a team of highly qualified experts and is a driving force behind the Cairngorms Connect partnership. And while I've not met him, I have enjoyed the privilege of a day out on the land with those experts from all the partners, and that night I slept well with a smile on my face. Not due to time slogging on the hill, but due to their inspirational company and through bearing witness to the fruits of their imagination and hard work.

Frankly it's astonishing. And that's just the map spread on the Land Rover bonnet. Cairngorms Connect is big; as big as we've got anywhere, and it's getting bigger. There's one significant neighbour that needs to get its act in order and get into the project, but even now it's impressive. Of course, it's got truly iconic landscapes as the fundamental part of its fabric and some command a real sense of unspoiled wilderness, but again what impresses me most is the project's vision. This cannot be a quick fix. You can't rock up and plant a granny pine; you have to wait for it to grow. So for now it's about nurturing seedlings, about having the courage to be shaping something you will never see finished in your lifetime.

We begin on the floodplain of the River Spey, on one of the RSPB bits – the Insh Marshes – where allowing winter flooding has created a riparian wonderland of mires, pools and fenland. It's one of the most important wetland reserves in Europe and home to wintering whooper swans and breeding waterfowl and waders, including wigeon, shoveller, goldeneye, common snipe, lapwing, redshank and curlew. Next, we climb up through the Wildland estate, pausing to admire the birch and Scots pine forest regeneration facilitated by the strict management of deer numbers. Here the trees speckle the glen; they are not in blocks or regimented lines and the land begins to look more primal. A quick lunch in one of the most picturesque bothies in the world, set on the banks of a river tearing its own unmanicured course through the heather, and then on to the Scottish Natural Heritage land, where we discuss the increasingly favourable responses to the grand plan that has arisen from the scientists, stalkers, farmers and foresters, and how together they will form part of a productive economic model in a landscape full of business opportunities, as the adventurers, photographers and hunters come to savour this spectacular place. And yes, I said hunters. Sustainable shooting, culling in the absence of those large predators, will avoid the Oostvarderplassen effect, and strictly regulated sports shooting is part of the plan. There's no driven grouse shooting here, with all its associated horrors, but low-impact walked-up grouse shooting is practised under the admiring eye of eagles and hen harriers. Finally, we stand before a massive 'clear fell' where Forestry and Land Scotland have laid waste to an enormous and sterile swathe of non-native lodgepole pine. Such devastation is never normally a public vote-winner, and many people still see any tree felling as environmentally destructive, but here ongoing effective communication and education resulted in negligible complaints and already the heather is sprouting and birch budding after years in the cold repressive shade of tree plantation.

The whole experience is heartening. These passionate people playing to their strengths, unshackled by the privateer, involving and benefiting local communities, ambitiously entwining their futures with the land and its wildlife, informing, educating, changing. And ultimately winning. Yes, it might be a 200-year journey, but already it holds 50 per cent of the UK's capercaillie, eleven species of regularly breeding raptors, over 5,000 species of plants, animals and fungi, 20 per cent of which are nationally rare or scarce, all secure among 10,000 hectares of peat-rich bogs, 13,000 hectares of native woodland and the biggest single remnant of ancient Caledonian pinewoods. Bloody hell, it's good. Get on the train!

RECLAIMING THE SEAS

When you imagine the word 'rewilding,' I bet you that the image in your mind is that of forests, peatland, bogs or moorland. The oceans are often forgotten or pushed aside – out of sight and out of mind. Perhaps because only three million out of the 60 million people who live in the UK are situated on or near the coast; those who live further afield are less connected and are therefore largely unaware of the state of our underwater systems. But less than 15 per cent of the European coastline is now considered healthy and viable, and on a global scale only thirteen per cent of the sea is deemed truly wild, with only 5 per cent of the entire ocean protected under legislation. Five per cent. That's it. It's nothing when you consider how fundamentally important it is to all life on this planet; producing somewhere between 50 and 80 per cent of atmospheric oxygen, mitigating the climate, regulating weather systems and providing food. It's not that marine restoration isn't taking place, because in fact it is happening all around the world, but it's considerably less advanced and is rarely discussed to the same degree. So let's look at a couple of examples ...

Sometimes the most effective rewilding strategy is to simply allow nature to do its thing. Bikini Atoll is the Pacific island where 23 atomic bombs were dropped by the United States during the Cold War between 1946 and 1958. One device, set off in 1954, was 1,100 times larger than that of the Hiroshima atomic bomb. It wasn't until 2017 that scientists from Stanford University went back to the site to investigate the impacts of the radiation and to discover if life had persisted. With reports of mutant sharks missing dorsal fins and a real-life SpongeBob SquarePants, it's no wonder this nuclear wasteland attracted some curiosity, but what they found was more than just a pineapple under the sea. Seventy years after the bombings, life was thriving. Corals the size of cars, hundreds of fish schools (snapper, tuna and sharks – all good indicators of a healthy environment), and coconut crabs. Professor Stephen Palumbi described it as 'quite odd' and 'remarkably resilient'. This was a case of extreme violence where people were forced out and as a result unintentional rewilding took place, but can people and deliberate oceanic rewilding coexist together?

Marine Protected Areas (MPAs) and ocean sanctuaries have been shown to boost biodiversity in their localised areas and facilitate the regeneration of habitats after intense human-induced damage. In places where marine ecosystems are intertwined with communities, it is possible, and sometimes necessary, to aid in the regeneration of habitats and species by creating legislation and lending a helping hand. In 1984 Sir David Bellamy founded the St Abbs & Eyemouth Voluntary Marine Reserve, the oldest of its kind in the UK and now famed for its clear waters and spectacular biodiversity. Both St Abbs and Eyemouth were originally fishing settlements, relying heavily on the productivity of the sea. Watching as the waters became increasingly polluted and overfished, the community of fisherman, scuba-divers, conservationists and swimmers came together to regulate the waters. Today, due to their efforts, it's a top spot for wildlife and supports a sustainable offshore fishery. Our approach to oceanic rewilding must be dynamic and innovative. I don't have all the answers, but adaptive methods on community-driven projects seem like a great way forward.

Is it a ghost? No, it's a stork!

My favourite film as a child was Dumbo. *I would make Chris watch it on repeat with me. I'd laugh, cry and hide in fear as the psychedelic elephants danced across the screen, but the very first scene is the one that I remember most fondly. White storks gliding in en masse with bundles of cloth in their bills presenting baby animals to their parents. These birds have such a strong cultural significance around the globe, associated with life in its purest form.*

The origin of this myth is hard to pinpoint as it is one of the most universal folktales, with variations embedded in European, African, Middle Eastern and American history. Some of these tales can be traced to ancient Greece and the story of the goddess Hera. Hera was immensely jealous of the beauty and power of the queen, Gerana. In anger, Hera transformed the queen into a stork. Heartbroken, Gerana desperately wanted to find her young child, and when located she flew off into the horizon with her baby clutched in her beak. Although very early depictions of this story are in fact centred around cranes and not storks, it appears the correct species this folklore is based on might have got lost over time or translation.

Another tale originated in Germany and Norway during the pagan era more than 600 years ago, when couples would marry during the summer solstice as it was believed that summer and good fertility were intrinsically interlinked. In the days after solstice the storks would leave on their annual migration, returning nine months later just when the newlywed couples might be welcoming their new arrivals. I just love the romantic element of that story – that the storks return, bringing in a new wave of life.

Their relevance to UK culture is also embedded in our street and village names. Storrington was called Estorchestone during the Anglo-Saxon period and means 'the village of storks'. So it seems fitting that efforts towards white stork conservation are centred around the Knepp Estate, which happens to be just around the corner from the former Estorchestone ...

White storks have not bred in the UK for 606 years – until now. Where there are progressive and positive rewilding schemes, there are opportunities for some very exciting reintroductions. Archaeological evidence tells us that this species had been breeding here from the Pleistocene era, some 360,000 years ago, up until 1416 when a pair nested on the roof of St Giles' Cathedral in Edinburgh. Stork bones have occasionally been found around the nation, from the Isles of Scilly in the very south of England to the Shetland Islands in northern Scotland. They did breed successfully here for thousands of years, so what on Earth happened to these long-legged beauties?

Storks have a very strong degree of natal philopatry, meaning that adults will return to breed in the same area that they were bred themselves. This predictability and the fact they always nest close to human settlements means they've been an easy target. Their persecution was widespread; and during the English Civil War (1642–1651) they were shot in huge numbers because of their association with the rebellion. White storks went from being relatively common to incredibly rare pretty quickly. In the years that followed, we wrecked their habitats by draining wetlands and fragmenting more of their landscape. The decline in white storks is hardly surprising given these pressures. Each year only about 20 migrant white storks are spotted around the UK, but due to their breeding requirements it was unlikely they would ever naturally recolonise. That's when Project Stork stepped in.

Project Stork's overarching ambition is to restore a population of 50 breeding pairs to the UK by 2030. No small feat! Following in the footsteps of successful white stork projects elsewhere in Europe, a total of 194 juvenile birds from Poland and northern France have so far been brought to three sites across the south-east of England. The birds were all wild, but had sadly sustained serious injuries that required rehabilitation over a period of time, so they were relocated to pens at Wadhurst Park, Wintershall Estate and the Knepp Estate. Once healthy and able, the storks are released, and many can already be seen free-flying close by. The hope is to establish a breeding population of 20 pairs at each location. These birds have been supplemented by releases of fledglings bred at Cotswold Wildlife Park, with the first 24 birds released in 2019. It was big news in May 2019 when one pair at Knepp built a nest and the female laid three eggs. The first of their kind to breed successfully since 1416! Unfortunately, it wasn't to be that year as the eggs failed to hatch, probably due to the young age and inexperience of the parents. But this was far from bad news – it was progress, important progress.

In the spring of 2020, when we were isolating in our homes, three stork pairs at Knepp began busily building nests. Two out of the three managed to successfully lay eggs and on 6 May project officer Lucy Groves watched as one of the adults removed eggshell from its nest indicating the very first hatching! It just so happened to be the very same pair that had attempted to breed the previous year. On 9 July the first chick fledged the nest, shortly followed by three others days later. This is the first generation of white storks that will, hopefully, be soaring in our skies and breeding in our oak trees for years to come. In my opinion, this was one of the most exciting conservation successes to happen in years!

Chris and I were humbled when Project Stork asked if we would break the incredible news and follow the storks' progress on our Self-Isolating Bird Club broadcasts. We were both genuinely SO excited to watch over these storks remotely with regular updates coming in from Lucy. In our broadcasts, we got to discuss some great new science and stories about the natural world, but this was the one that took the crown – the top spot. The four white stork chicks are testament to the brilliant rewilding work at Knepp and the dedication of Lucy and her wonderful team of volunteers at Project Stork. Who knows how many there could be next year!

BEAUTIFUL ZOMBIES

Now you don't see it, now you do. It's magic. And we love magic, admire trickery – the empty hat is suddenly filled with a fluffy white bunny, or better still the fluffy white clouds are suddenly filled with red kites! But that's not magic or trickery, that's just some of the best and most creative conservation we can muster – not that we've always pulled the right species out of the topper.

The UK is not a pristine utopia – hardly breaking news – in fact it's a pretty ghastly mix we've picked by importing species from all over the world while carelessly exterminating those natives which really should be here. The result is a confused and often conflicted ecology where the so-called balance of nature lurches wildly and can be unhealthy, even lethal, and thus both environmentally and economically catastrophic. In short, there's many a man-made mess out there in the species mix we've concocted and, again, that's why I think we sometimes make ourselves feel so good when we achieve something bold, like a reintroduction, and see it as a quick fix. The question is, have those beautiful swirling cyclones of red kites actually fixed anything in our rotting landscape, or are they just a vanity badge that makes us feel good? Which may beg another question – are reintroductions really worth it? And for that matter, is the over-zealous removal of non-natives really worth it? Is the cash calming the chaos? Does the economic input equate to the ecological recovery? Let's look at some of our favourite loves and loathes.

There are a great many to choose from across the flora and fauna, but let's start with those old foes: grey squirrels and American mink ... *grrr*, I can feel blood pressures rising. Keeping it super simple, the problem with grey squirrels is that they spread a disease that has decimated our native reds, and the riverine mustelids, American mink, have munched many a local water vole. The latest Mammal Society listings, sanctioned by the International Union for Conservation of Nature (IUCN), categorised both the red squirrel and the water vole as 'Endangered' (and, as a nasty aside, reveal that one in four of our native mammals is threatened with extinction). So, we despise the invaders, kill them with impunity – there is no closed season – and use them as zoological counter parts to bolster our own horrible xenophobia. Yes, a scientific appraisal of several notable media outlets found an ugly correlation between the apparency of stories about grey squirrels and human migrants. Some people, eh? Anyway, back to the ecology, and there is no ambiguity – the greys and the mink are enormously destructive and an undesirable addition to our fauna. The rodents arrived in 1876 – though the mink weren't breeding in the wild until 1956 – and they have both proliferated because they've squeezed into vacant or modified niches, and now cover most of the UK. The exceptions for one or the other of the species are the Western Isles, the Orkney and

Shetland archipelagos, the Isle of Wight, Anglesey and parts of northern Scotland. In short, they are 'everywhere' and often abundant, which means getting rid of them would be very, very difficult and expensive, but not impossible. And thus so far we have targeted our removal programmes to protect local populations of reds and water voles with therefore limited success. A plan to trap grey squirrels to form a 'no-man's-land' across the north of England or southern Scotland to protect the reds' highland stronghold might have been some sort of option if some idiots hadn't released greys into Aberdeenshire in the 1970s. An initiative to produce a vaccine for the squirrel pox disease has been explored, but couldn't find sufficient funding and became mired in petty infighting so, currently, as per usual we seem to be sitting back and waiting for the *Sciurus vulgaris* shit to hit the fan before we deal with the scourge of *Sciurus carolinensis*. We moan about it, but don't fix it.

Not so in New Zealand, home to the Noxious Animals Act of 1956. No, antipodean conservation means business, and it's getting on with the job of dealing with its hugely destructive non-natives with a government co-funded joint venture company, Predator Free 2050. And guess what, they've spent some money – NZ$ 35 million from the government, plus extra funding whereby for every $2 donated by the public, philanthropists or local councils, they give an extra $1. The targets are possums, rats and stoats, and, er, feral cats. Issues about the latter have predictably courted controversy, but the initiative has been very popular as has 'backyard trapping', whereby individuals and communities play a vital role not only in removing animals, but critically in education and generating momentum. In Wellington alone there are now more than fifty groups on the prowl for pesky predators. Imagine if our governments would commit some serious finance and training to such schemes ... Not so I'm afraid, and its neglect is costing us dearly.

In 2010 stoats were spotted on Orkney for the first time. A native predator to the UK mainland, but one that was either deliberately or accidentally translocated to a place where they had never lived. Consequently, they posed an immediate threat to the unique Orkney voles and ground-nesting birds, such as hen harriers, short-eared owls and a host of waders and wildfowl, for which these islands were a secure stronghold. Instant action was required ... so four years later in 2014 the Scottish government's statutory agency Scottish Natural Heritage commissioned ... a ... study. Guess what the conclusions were in 2015? Yep, that stoats posed an immediate threat to the unique Orkney voles and ground-nesting birds such as hen harriers, short-eared owls and a host of waders and wildfowl for which these islands were a secure stronghold. Wow. After a trial run in December 2017 – yes, a trial – eventually in August 2019 the first traps were laid as part of the Orkney Native Wildlife Project at a cost of £6.1 million of taxpayers' money over five years. In the nine years between their arrival and the traps being set, the stoats of course had become fully established and widely distributed throughout mainland Orkney, Burray and South Ronaldsay. A costly case of closing the door after the ermine had bolted. And just when we thought things couldn't get worse, the Scottish Gamekeepers Association piped up with the idea that the wrong-sized traps were being used, animal welfare being one of their primary concerns. Needless to say, it came to nothing, as did a short-lived #SaveOurStoats campaign. As I write stoats are still at large on Orkney, presumably in large numbers, chewing up rare birds with quicksilver abandon.

Okay, so trying to get rid of entrenched non-native invasive species is hard, or we make it hard, but it's also very expensive. In 2010 CABI reported that it was costing the UK economy £1.7 billion a year. The report says, 'Plants, mammals and plant pathogens are the organism groups that cause the largest costs. On a species level, rabbit and Japanese knotweed cause most costs, followed by brown rat'. That presumably will have changed, given the decline in rabbits due to the haemorrhagic disease that has wiped them out of large areas of the UK – good news for the economy, bad news for stoats (not on Orkney), buzzards, kites, foxes, etc, etc. And getting rid of brown rats is also causing problems through the use of second-generation anticoagulant rodenticides (SGARs) such as Neosorexa and Slaymor. The Barn Owl Trust have been constructively vocal on this issue. Their data indicates that 100 per cent of kestrels, 94 per cent of red kites and between 78 and 94 per cent of barn owls were contaminated with these SGARs, and even 57 per cent of hedgehogs had traces in their bodies. The big question is, of course, are these sub-lethal doses having any ill effects? The pesticide brigade are quick to point out there is no evidence to suggest they are; I'd be quicker to point out that there is no evidence to suggest that they are not. We know that safe medicinal doses of the anti-coagulant Warfarin can have side-effects in humans, and that the poisons found in barn owls are 100 to 1,000 times more toxic than Warfarin. Even if behavioural changes in these predators are minor, with 76 per cent of farms using highly toxic rodenticides and with so many of our predators affected, the overall effect may be significant because almost the entire population is exposed.

Many European countries are opposed to the use of SGARs, and in 2012 the prospect of an EU-wide ban upset the rodent control poison cart – mainly because there were no new chemicals coming online. Look, brown rats can be a threat to human health through food contamination, and their presence on offshore islands has devastating consequences for nesting seabirds, but while it is an inescapable fact that rats need to be controlled, poisoning may *not* be the best way to deal with the problem here in the UK. In fact, in some parts of the UK there is a rising resistance to these poisons in rats, meaning that users are simply throwing more poisons at their 'problem' – and our wildlife. That said, worryingly, the use of poisons is set to continue and so far use-restriction regulations and statutory safety notices on SGARs have failed to prevent wildlife contamination.

From 2015 to 2017 a voluntary stewardship scheme was phased in whereby large quantities can only be sold to those farmers, gamekeepers and professional pest controllers who can prove that they have attended an approved training course. But the scheme is being led by a group managed by the rodenticide manufacturing companies ... Meanwhile, dead barn owls collected by the public in 2019 and analysed showed that 87 per cent contained one or more SGARs compared to a pre-stewardship (2006–2012) baseline figure of 81 per cent. Clearly, the requirement for rodenticide user training has so far proved ineffective. What a surprise.

My point is simple, in that it is not simple. When it comes to dealing with, killing, managing or eradicating the plethora of non-native invasive species we've deliberately or accidentally let loose on the UK's landscape, the issue is difficult ethically, practically, ecologically and economically. In short, the next time you curse that grey squirrel gnawing your bird feeder, the knotweed or balsam messing with your lawn or streamside, or the harlequin ladybirds hibernating in your window frame, ask yourself: what good would it really do to kill them, at what cost and what it would realistically achieve? And let me be clear that I confront that question often – most mornings before the poodles go out on patrol the greys are pilfering my offerings to the goldfinches, greenfinches, tits, et al ... and although I huff and puff, and count the number of Sid and Nancy's unsuccessful hunts, in truth I've made peace with the raiders. But if, on the other hand, our government or its statutory bodies were to wake up to a serious nationwide plan to emulate New Zealand, I might for a moment think again. Just long enough for me to remember the ruddy duck fiasco – but that's another story. Let's move on to putting things back rather than taking them out.

In the UK, we could justifiably claim to have a rich tradition of successes when it comes to reintroductions, and we got off to an early start with the capercaillie, which had completely disappeared by the mid-eighteenth century due to habitat loss and overhunting. Birds from Sweden were reintroduced into Perthshire in 1837 and by the 1970s there were about 20,000 of these majestic and charismatic giant grouse in the Scottish pine forests. In recent years they have subsequently declined again to a population of around 2,000, and in truth they are now hanging on at the edge of their natural range in impoverished habitats and subjected to an increasingly unsuitable maritime climate of wet springs and summers. Of course, there are on-going attempts to save the capercaillie by promoting woodland regeneration and blaeberry abundance through deer management, and eliminating predation by pine martens – another idea favoured by the Scottish Gamekeepers Association. What they fail to understand, however, is that ground-nesting birds such as this have evolved to live with predators and do so successfully elsewhere in the world. Eliminating one species to save another is not a sensible long-term solution. The answer lies with large-scale habitat restoration, namely projects like Cairngorms Connect, which, as we saw in the previous chapter, is now home to 50 per cent of our capercaillies, and plenty of pine martens too.

BRING BACK THE BISON

For the first time in 6,000 years, European bison are planned to return to the wild in the UK. In 2022 a small herd is set to be released in Kent as part of a £1 million project to safeguard the future of the endangered species and to help promote national biodiversity and ecological restoration. It is part of a wider reintroduction project taking place in different countries across Europe, to help save these megafauna giants that can weigh over 600 kilograms.

European bison are the closest living relatives to the steppe bison, which used to roam throughout the UK but became globally extinct approximately 10,000 years ago at the end of the last ice age. Some of the most famous cave paintings, in Lascaux and Chauvet-Pont d'Arc in France, depict steppe bison among other species, but there is one drawing that for the longest time puzzled researchers as it neither looked like a steppe bison nor the existing European bison we have today. It turns out this wasn't a badly drawn illustration, but in fact was a clue about how European bison evolved. Scientists used ancient DNA samples taken from radiocarbon-dated teeth and bone found in caves throughout Europe and discovered that this unusual-looking bison painting was a crossbreed between steppe bison and the ancestors of modern cattle, the aurochs. Prior to this discovery, no one really knew how European bison evolved as they seemed to just 'appear out of nowhere' in the fossil record, but now it's believed they are descended from the crossbreed depicted on the cave more than 20,000 years ago. A big surprise for scientists as hybridisation leading to an entirely new species in mammals never really happens!

Bison are another example of ecosystem engineers (along with beavers) as they naturally regenerate old pine plantations by eating the pines' bark and rubbing themselves up against the trees in order to moult their thick winter coat. This kills the pine wood and makes way for a more biodiverse and healthy woodland. The dead wood created becomes great habitat for insects, which in turn will support a network of other animals, like birds and bats. Nightingales and turtle doves are among the species expected to really benefit from the presence of bison due to their engineering work. As it stands, just four bison – three females and one male – coming from either the Netherlands or Poland will be allowed to roam an area of 1,236 acres. The size of the herd will naturally increase with breeding, and once the herd is settled, the public will be able to go and visit these ginormous docile mammals. Stan Smith from Kent Wildlife Trust said, 'sometimes in the rewilding debate people think that it's a look back to the past, but that's not what we're about. We're about trying to find the right natural solution for the modern world.'

I've seen large copper butterflies in Europe – they are an absurdly striking insect, very bling, a bit like a scarlet Lamborghini Aventador in that they are embarrassingly overblown, but as much as you want to snort with derision you just can't resist staring at them. As they spark across the marshlands they leave every other lepidopteran trailing in their spectacular wake and looking pretty ordinary. Broom, broom!

In the UK they were first described from a specimen taken at Dozen's Bank near Spalding in Lincolnshire in 1749, but were probably last seen flashing their brilliant wings just 102 years later in 1851. There are some vague records going on until 1864, but at this point the precise date of their sad demise is all but immaterial. Crazed Victorian collectors mopped up the last few but by then the fens had changed; they had been drained dry in the relentless clamour to plough and plant their rich black sod with crops. The insect's incomparable allure has meant that there have been several reintroduction attempts, the first and most enduring at Woodwalton Fen in 1927. This sort-of worked, but the population had to be rigorously pampered and repeatedly reintroduced or supplemented from captive-bred animals from a rare Dutch sub-species, and ultimately – and unfortunately – all reintroduction attempts have failed and there none currently at large in the UK. Bah! Boo! Damn! And why not?

Research conducted by Andrew Pullin at the University of Keele in the early 1990s discovered that these attempts at reintroduction appear to have foundered during the overwintering stage of the butterfly's lifecycle, after which up to 95 per cent of the caterpillars inexplicably failed to reappear. The larvae hibernate among the dead leaves of the greater water dock, which wilt and fall onto the ground in autumn. Their plan is to remain close to the parent plant so that when its leaves sprout in the spring the caterpillars crawl up the stems to an instant source of the only type of food that they eat. But the curious thing is that greater water dock is common throughout the UK, so the larvae don't perish for lack of food. It's not the cold that kills them either – Pullin established that the caterpillars can survive temperatures as low as -25 degrees Celsius and can also survive periods underwater, so flooding also shouldn't be a problem. One possibility behind the failures he considered is that there could be excessive populations of predators such as spiders and birds. Another is that the number of insects released in the past has been too small and the colonies too isolated. We know that some butterflies, such as marsh fritillaries, can only survive in what we call metapopulations – essentially a population of populations, or a group of groups, that is made up of the same species. Each subpopulation is geographically separated from all the other subpopulations, but movement of individuals from one to another occurs regularly. When there are not sufficient subpopulations sufficiently close together they go into terminal decline and that metapopulation becomes extinct. It may well have been that the core area, Woodwalton Fen, was just too small, which is interesting as this is no longer the case.

The Great Fen project should have had some praise in the previous chapter as it's an exciting and ambitious 50-year plan to create a huge wetland area comprising some fourteen square miles between Peterborough and Huntingdon, and is destined to be one of the largest wetland rewilding projects of its type in Europe. It encompasses both Holme Fen and Woodwalton Fen, two of the large copper's historic haunts. So maybe there is still hope, and a renewed interest in a fully-formed and better-informed reintroduction could see this sparkling gem flitting in our summer sun once more. After all, we have worked wonders with other extinct butterflies.

I've never seen a large blue so I cannot eulogise about their apparent splendour and compare them to a supercar, but they became extinct as The Clash were recording 'London Calling' and were successfully reintroduced in time for The Smiths (more their sort of thing), so that's got to be worth some consideration and subsequent celebration.

In a nutshell: by the early 1800s the species was rare in the UK and is now endangered globally, t was being intensively studied before disappearing in 1979, its complicated life history was finally understood, habitat management was organised accordingly, and it was put back using Swedish stock in 1984. Fast-forward to 2016, by which time there are 10,000 on the reserves in Somerset and Gloucestershire alone and the UK has the highest concentration anywhere in the world. Job done – the project masterminds, Jeremy Thomas, David Simcox, Judith Wardlaw and Ralph Clarke, should have got knighthoods. They didn't, but they are up there with conservation's greats.

In truth, back in 1979, they were about to save it before they lost it. Rabbits cost them the game, or rather the lack of rabbits. Like I said, it's complicated, very complicated – but stay with me. The large blue has a very peculiar lifestyle and thus very precise requirements. When it hatches in July its caterpillars feed on the flower buds of wild thyme plants for three weeks and then drop to the ground, where they so superbly mimic the smells and sounds of the red ant *Myrmica sabuleti* that they are carried away to the nest and nurtured as if they were actually ants themselves. Then they spend the next ten months in this sheltered environment, and their mimicry ensures that the ants leave them alone even when they eat the ants' larvae. You couldn't make it up! A carnivorous butterfly that is dependent on an ant! Brilliant, but like so many precise partnerships, it's so precarious.

The large blue's problems started when subtle changes in grazing and local vegetation caused this particular type of ant, to be replaced by other species that were less suitable surrogate parents. Why? Well, it turns out that *M. sabuleti* is very sensitive to the height of the vegetation surrounding its nest. If it grows taller than 1.4 centimetres the overgrown plants reduce the temperature of the earth where the ants' brood-chambers are located. They are then soon outcompeted by other ants that are less dependent on warm soil and their numbers fall. Cue the rabbits, which nibble a close sward and keep the ants cosy. Except that in the 1950s myxomatosis was introduced and it killed off the rabbits on the hillsides where the *sabuleti* ants lived. The turf grew longer, the ants disappeared and with only second-rate babysitters the large blues went into terminal decline. Extremes of wet and dry weather through the early seventies piled on the pressure and the last few active butterfly collectors nabbed a few from the remaining two colonies. And we only know this due to Thomas, Simcox and Clarke's comprehensive research. Using their data, they created a mathematical model to map all the factors that influence the population size of the butterflies and inform their evidence-based conservation. At its core *M. sabuleti* has to be close to wild thyme plants, so scrub management and grassland restoration, along with sheep, cattle or pony grazing, are critical to keeping conditions right. Next year, if lockdown allows, I really must go and see a large blue, not just to satisfy my entomological cravings but to honour the brilliant work of all those conservationists who have made that aspiration a reality.

BACK FROM THE BRINK

Back from the Brink (BftB) is one of the most ambitious conservation projects in England, organised by Natural England and Rethink Nature and funded by the National Lottery. When it comes to reintroducing wildlife and protecting its habitats, there needs to be joined-up, collective thinking and that is exactly what BftB is. Rethink Nature was set up by seven different organisations: Amphibian and Reptile Conservation Trust, Bat Conservation Trust, Bumblebee Conservation Trust, Buglife,

Butterfly Conservation, Plantlife and the RSPB. And with their combined support, BftB aims to save 20 species from the brink of extinction while benefiting over 200 more indirectly in nineteen different projects throughout England, spanning from Northumberland to the south of Cornwall. Pine martens, field crickets, willow tits, black-tailed godwits, lesser butterfly orchid, natterjack toads and Cornish path moss are just some of the species benefiting from their work.

The little whirlpool ramshorn snail is one of the rarest freshwater molluscs in the UK. It is not normally bigger than five millimetres and relies on chalk waters, like marshland ditches, that are unpolluted, clear, full of vegetation and have a high pH. This habitat is so increasingly rare itself that this little snail is now restricted to just three sites in south-east England. Carp are also an issue as they stir up the sediment in suitable places, which then makes the water unsuitable for the snails. The RSPB are overseeing the project to help restore the population by managing ditches and creating more in the hope that the species can be reintroduced into the RSPB Pulborough Brooks reserve where it was once present. Another of their projects is to maintain the Cotswolds' unimproved limestone grassland – unimproved meaning that it has never been sprayed or ploughed. Over 50 per cent of the country's unimproved grassland occurs in this area, but more than 95 per cent of it has been lost since the 1930s. This project, fronted by Butterfly Conservation, works with landowners to better manage and monitor the habitat for the benefit of many species, such as the red-shanked carder bee, fly orchids, basil thyme and the rare greater horseshoe bat. Beginning in 2017, Back from the Brink is still in its early days in the grand scheme of things, but their projects are making serious headway. It's their combined thinking and collaboration that will save these species – I take my hat off to them.

The large blue's reintroduction was complex and required enormous human input, but another one of our most famous returns was a result of us simply not doing something – namely, killing. Ospreys were once widespread all over the UK, as they are over much of the rest of the world, but due to unimaginable levels of persecution at medieval fishponds, and then from trophy collectors and by gamekeepers, which reached apocalyptic levels in the 1800s, they were extinct in England by 1840 and Scotland by 1916. They continued to appear annually as a passage migrant and occasional breeder, until 1954 when birds, probably of Scandinavian origin, settled in Strathspey. In the early years egg collectors and vandals were an issue, but by 1959 a pair had settled at Loch Garten RSPB reserve where they have continued to breed almost every year since. The population was slow to grow at first; by 1976 there were only fourteen pairs, by 1991 it had grown to 71 pairs. Then things took off – by 2001 there were 158 and that same year, through natural recolonisation, they successfully bred in the Lake District. Through reintroduction by the Roy Dennis Wildlife Foundation and Tim Appleton at Rutland Water, the first young fledged in England for 160 years. In 2004 two pairs nested in Wales, the males being birds that had been translocated from Scotland to Rutland. There are now well over 300 pairs in the UK, most in Scotland, but we were so, so close to having the first breeding in the south of England in 2020 thanks to another regional reintroduction by the Birds of Poole Harbour and the Roy Dennis Wildlife Foundation.

Roy Dennis is right up there – one of the very greatest modern conservationists, a legend – particularly when it comes to ospreys. He was in at the start, working alongside George Waterston et al in the Abernethy Forest from 1960, and also with John Love et al on the reintroduction of white-tailed eagles on Rum in the mid-1970s. In black and white photos from those pioneering days he cuts a dashing figure; I've called him the 'Clint Eastwood of conservation', not that he's keen on the epithet – probably because the last thing he would say is 'go ahead, make my day'. No, Roy is less sudden-impact, more imbued with a steadfast and resolute determination, a fairly democratic tongue and a rare degree of confident savviness, which gets him results. And at 80 he is still getting results. His foundation is now also reintroducing sea eagles on the Isle of Wight and last summer I went to see the six young birds that had been translocated from Scotland where there are now 130 breeding pairs. It was a sunny evening and I crept silently up to the release aviaries and squinted in. Big, brown and shaggy, already armed with a massive beak, and beautiful chestnut eyes frowning hard. There were six, though we are now down to four – one died, one is missing, two still on the Isle of Wight, but as their satellite trackers show us, roaming widely when they get the urge. It was a tremendous privilege peeping between the panels at these stunning predators, but the highlight of my evening was meeting Roy and revelling in his wisdom and enthusiasm, watching his eyes sparkle more like an eighteen- than an eighty-year-old.

Roy's foundation has a licence to release 54 more birds over the next four years, and they will. Covid-19 has stopped a lot this year, but no pandemic could thwart Roy's progress and eight more are being released. The last pair bred here on Culver Cliff in 1740 and, given that young eagles don't breed until they are about five, I can't wait until 2025. The hope is that a small population of six to eight pairs will be established on the island and within the Solent area, and then spread along the coast while connecting with the populations in France and the Netherlands. And it won't just be you, me, Roy and all the conservationists who are excited by this, because in Scotland eagle tourism is incredibly popular and lucrative. Recent analysis has shown that white-tailed eagles generate up to £5 million per annum for the communities on Mull and £2.4 million on Skye.

When it comes to other raptors, goshawks seemed to recolonise naturally after being persecuted to extinction in the 1880s, although escaped falconers' birds and a few 'accidental' escapes may have helped. There has been no census, and they'd be tricky to count as they're so secretive, but the consensus is that there are now between 500 and 600 pairs. Unfortunately persecution means they fail to re-establish in areas where there is game-rearing interest and, worse, spread to areas where there is not. Urban Berlin has reached carrying capacity; no more goshawks can be squeezed in, as over 100 pairs feed in the parks and gardens on pigeons, crows, magpies, squirrels and rats. I know what you're thinking – yes, if they were allowed to reach our cities they'd be predating all those species most people don't like. Stop it!

The red kite reintroduction has been so well documented and so deservedly celebrated elsewhere that I'll leave you to research its recent history and marvel at its widespread success, because wherever you are in the UK there is a chance that you might look up and see a red kite. I once had a meeting with Michael Gove, then-Environment Secretary, and when I walked into his office in the heart of Westminster I spotted two out of the window, which we took time to enjoy. Anywhere goes! You see, 30 years after nineteen birds were flown in from Spain to the Chilterns there are now nearly 2,000 breeding pairs and 10,000 birds in total, which is, get this, 10 per cent of the world's population. They still get shot and poisoned by a few criminal idiots, but this surely has to be UK conservation's greatest single species success story to date. Not that our reputation is entirely intact, because that reputation is challenged by one contemporary colossal and embarrassing failure. It's the very sore open wound that is our wholesale failure to reintroduce the beaver.

It's not our fault, not the conservationists' fault. Every last one of us now knows and fully appreciates the long list of clear advantages this reintroduction would have, all of which have been demonstrated and measured across Europe and the world, and now through successive successful trials here at home. It's utterly exasperating to see this very necessary job constantly put on hold. The various governments' statutory bodies, Natural England, Natural Resources Wales and Scottish Natural Heritage, have all behaved very badly, particularly the latter. But the real monsters in this mess are the so called farmers' and landowners' unions. They have manifested unbelievable ignorance and antipathy, the result of which has led to the inhumane and unnecessary slaughter of the beavers that are currently living in the UK's waterways. Even though (another) scientific trial of beavers living freely on the River Otter in Devon has (again) concluded that they deliver enormous benefits, not just for biodiversity, but flood management, water quality and tourism, the government has yet to authorise their return. A decision on beavers in England had been due by the end of summer 2020, but Natural England have said they couldn't give a timetable because of delays caused by the coronavirus crisis. It's deplorable.

FOR THE LOVE OF BEAVERS

Beavers are iconic. Our landscape would never have developed the way it has without them, and yet we rid that landscape from them nearly 500 years ago, and we're still suffering the consequences of that void today. These semi-aquatic natural engineers have been shown to reduce peak flooding rates by over 30 per cent, and they can rapidly reshape the landscape, creating wetland habitats that support a wealth of biodiversity – and all because of their love of woody plants.There has been lots of research into the benefits of beaver reintroductions – too much for their release to even be questioned – but there are also some interesting studies revealing insights into the evolution of their woodcutting and dam-building behaviours.

Beavers belong to the *Castoridae* family, a group of herbivorous rodents that evolved during the late Eocene period. The very first definitive indication of woodcutting beavers was found at the High Arctic Beaver Pond fossil site dated back to the Pliocene, although evidence suggests that this beaver behaviour originated between 20 and 24 million years ago. One of the first species would have been the now-extinct giant beaver that could grow up to 2.2 metres long.

Despite their size, the incisor marks left behind on the preserved wood suggest beavers were not so well equipped at woodcutting then as they are now. Researchers measured the stable carbon and nitrogen isotopes of early beaver collagen and subfossil plants to try to understand why this behaviour occurred in the first place. This showed that early beavers were dependent upon a diet of freshwater macrophytes (submerged or floating vegetation) and woody plants, suggesting that woodcutting solely evolved as a way of harvesting plants for consumption. Dam building came later, made possible by the result of their adaptations for swimming and woodcutting, and driven by the global cooling event during the late Neogene period. Dam building would have been an advantageous skill in a cool climate to maintain temperature. These events have led to the evolution of the keystone beaver species we have today, providing fundamental ecosystem services that are invaluable to the countryside ... or at least they would be if they were allowed to just do their thing!

After a struggle, we have great bustards back on Salisbury Plain and common cranes on various wetlands where introduced birds have, rather like the goshawks, supplemented a population that has naturally recolonised from the continent. Corncrakes have been returned to the Nene Washes, black grouse to the Peak District, water voles and dormice have been captive-bred and released all over the UK, initially with increasing and now with assured success. But to conclude we'd better get off the great and the good, and get down and dirty with a few of our other successful reintroductions, although they are more national relocations or local reintroductions. Notable results have been seen with both sand lizards and natterjack toads, which have been captive-bred or translocated to many new sites; pine martens have been recently released in Wales, where they have spread rapidly to expand a former population that was hovering at low to non-existent levels, and field and wart-biter crickets have been bred and released into areas of managed habitat. South East Water spent 20 years creating a home for the latter around Deep Dean Water Treatment Works in East Sussex, as these picky orthopterans need a particular pitch, which includes bare ground, short turf and taller clumps of grass in which they can hide from predators, notably magpies.

Plant-wise, more work needs to be done as success rates are low. Results indicate that survival, flowering and fruiting rates of reintroduced plants are on average 52 per cent, 19 per cent and 16 per cent respectively, and that an increased focus on species biology is required, a higher number of transplants is needed (preferably seedlings rather than seeds) and consistent long-term monitoring after reintroduction is necessary. There are a few exceptions – the stinking hawk's-beard, which resembles a dandelion and smells of bitter almonds, is now growing in large numbers at Rye Harbour Nature Reserve and great sundew, strapwort and the black poplar tree have also enjoyed some renewed success.

I'm afraid that this sad indictment of our lack of investment and commitment to re-establishing plants may be indicative of a contemporary ill – our collective obsession with celebrity. Eagles, ospreys, kites and storks are glamorous, but they don't build ecosystems, they are supported by them. So of course if, as in the case of ospreys and eagles and storks, there is still the relic of a functional ecosystem there that can support them, then let's go. It's good for conservation's morale and the local economies where they prosper. But where there is the wholesale collapse of our insect populations, butterflies, moths, beetles ... there, individual reintroductions can be triumphs, they can secure a single species' future, but they are not going to save the day. What we need are those ecosystem engineers, those species that shape and support an enrichment of wholesale biodiversity – animals like the beaver, forests like the oak. But what's interesting is that while you and I can't reintroduce big birds or mammals, we can plant plants, in our gardens and in community spaces. We can reintroduce plants and trees locally and, if enough of us do, we can start from the bottom up and install the fundamental building blocks of ecosystems, rather than put cherries on a cake that is stale.

Are trees the answer?

The notion that planting trees will save the world has been circulated far and wide over the last few years, with governments and big organisations pledging to fund huge reforestation projects in the hope of mediating the impacts of the climate crisis. It's become the fashion – a barometer to show just how green you really are by proudly announcing to the world the total number of trees you've planted, or intend to plant. In the latest election in 2019, the Conservatives pledged to plant 30 million trees by 2024, covering at least 30,000 hectares. As it stands, shockingly, they're falling short. Under their new countryside stewardship scheme, it's estimated that they will spend approximately £5.2 million on replanting during the financial year of 2020. This might seem like a lot but in reality it constitutes only 1,260 hectares, going against their pledge to plant at least 5,000 hectares a year. Even that's a minuscule area in comparison to the 30,000 hectares we were promised. And research estimates that each taxpayer contributes less than £2 annually to foresting. (But don't worry – all your hard-earned money is going towards HS2 instead!) So much more needs to be done, it's no wonder the NGOs are taking this work into their own hands – just to name a couple, by 2024 the Woodland Trust aims to have planted one tree for every person living within the UK (this is the biggest tree-planting scheme in our history) and The Nature Conservancy has a project to grow one billion trees around the globe.

But is tree-planting really all that it's hyped up to be? Currently only 13 per cent of the UK has tree cover, which stores 48 million tonnes of carbon within the woodland itself *and an additional 42 million tonnes in the soil and leaf litter below. These figures are negligible, and honestly shameful, when compared to other European countries. Finland has 72 per cent tree cover, Sweden has 68 per cent, Austria has 47 per cent, France has 36 per cent and Italy has 35 per cent. Planting trees could, in some areas, be the easiest and cheapest way to protect ourselves from the impacts of a warming climate. As trees grow and mature they act as a carbon sink by absorbing the atmospheric CO_2 that we pump out by burning fossil fuels. In 2019 some new research from the Swiss university, ETH Zürich, hit the headlines with some pretty extraordinary results: a worldwide planting programme could have the potential to remove two-thirds of all anthropogenic carbon emissions. Analysis showed that 11 per cent of the Earth's land surface could be suitable for tree-planting without any risk of encroaching on agricultural or developmental zones. This is an area the size of the US and China combined and would give space for the addition of 1.2 trillion tree saplings! This idea is gaining popularity among politicians – even the likes of President Trump (the man who pulled out of the Paris Agreement) is up for it – largely because, I imagine, it sounds like an easy fix. Plant trees → remove carbon → carry on as 'normal' → plant more trees → remove more carbon → continue to carry on ...*

But this approach has come under considerable scrutiny. No one disputes that planting trees is a viable tool to help us in the struggle against a changing climate, but we must also seek to solve the root of the problem. Planting trees is not, and never will be, a free pass to continue screwing the planet. You wouldn't use plasters to prevent wounds; you'd simply try to stop falling over in the first place. We need behavioural change and systematic change. We must look at planting trees alongside other natural climate solutions, like the restoration and protection of mangrove swamps, peat bogs and seabed grasses. For whatever incomprehensible reason, these natural carbon-storing environments are only allocated 2.5 per cent of our government's budget for carbon capture despite their massive significance. But, more on this to come.

Planting trees is a good concept, but there are some serious questions that we need to ask. What tree species are you planting and where exactly are they going? If the government's strategy is to plant 30,000 hectares of fast-growing conifers in a commercial plantation, then there is very little point. Although fast-growing non-native conifers may sequester more carbon quicker than slow-growing broadleaf trees, at least half of these non-native trees in the plantation will be cut down within their first fifteen years to be used in low-quality products, like fencing or pallets, or for burning. During the timber harvest for 2018, 56 per cent was sent for use in sawmills. Some of this wood, if used in construction, will lock in the carbon for decades in buildings, but when it is sent to paper mills the stored carbon is released from the wood back into the atmosphere within a matter of years. When you consider the carbon impact of removing these trees, then the whole process will in fact add more carbon than it will take out. Thomas Lancaster, who is head of the UK land policy for the RSPB, said, 'we should not be justifying non-native forestry on carbon grounds if it's not being used as a long-term carbon store.'

We also can't plant these forests, whether they be native broadleaf or commercial conifer plantations, on the cheapest land that's available, which is blanket bog in Scotland. This would be hugely damaging for biodiversity, but also, as peat bogs hold a lot of carbon, any degradation would release more into the atmosphere – very counterintuitive. Other areas that are being discussed for these projects are the shallow peat moorlands in western Scotland, parts of the Lake District, the Pennines and South and Mid Wales. Concerns have been raised by the RSPB as these shallow peat environments are invaluable to many rare species, like curlew, which would not survive well in these forested environments. In terms of the best option for wildlife and the climate, planting a slow-growing broadleaf woodland that has long-term carbon sequestration potential would be ideal. Long-term means that it must be left untouched, but this does pose a new problem that needs some attention. I'll give you the general gist now to spark a discussion, but won't go into too much detail. It is argued, and rightly so, that the UK should have space for fast-growing plantations that enable us to produce our own timber, because otherwise we will have no choice but to export from overseas and therefore massively increase our carbon footprint. Lancaster observes, 'it's clearly not just a question of more trees equals a safer climate. Trees in the wrong place could exacerbate climate change and biodiversity decline.'

The aim of planting 30,000 hectares of woodland by 2025 set by our government is one that I fear will not be met until 2030, let alone by 2025. As it stands, Scotland is leading the way, having planted over 80 per cent of all the trees that have so far been established as part of this initiative. Their target is to reforest 12,000 new hectares each year and, if it wasn't for Covid-19 and some very wet weather, they probably would have reached it. In 2020 they've planted 10,860 hectares, with 7,570 hectares being made up of commercial conifer plantations and 3,890 broadleaf hectares. I'd like to see a higher proportion of productive broadleaf forests for the sake of our climate and wildlife, but they weren't too far off hitting their goal. Ireland, Wales and England need to catch up and pull their weight, especially in England where the UK government has direct responsibility. Globally, this is no small undertaking, with scientists suggesting we must reforest an area the size of one to two billion hectares. This could take anywhere between one and two thousand years to establish, assuming that we plant at least one million hectares a year containing somewhere between 50 and 100 trees per hectare. It's a rather big plaster for a rather big wound, and when it comes down to it, we really need to cut carbon emissions at the source by becoming individually responsible for the waste we pump out. However, trees could help soften the blow.

So, where to start? Well, I suggest one tree at a time because society grows great when old men (or women, or children) plant trees whose shade they know they will never sit in.

WHAT ON EARTH?

I'm in the light factory. In that massive machine that turns sunshine into oxygen and carbon, that almost miraculous and perfect invention that nature has evolved to make life viable and sustainable on Earth. It's so green it hurts, it's so loud I cannot be outside its influence, it smells like nothing else I know and being there has been a saviour through this Covid nightmare. I sit humbled beneath its apparatus, an irrelevant moment in its staggering existence; I'll be gone before I'm even a johnny-come-lately. Beneath these history generators I wonder and wish, crave to read their memories. Under this beech tree people have laughed and cried, been conceived and died, they have kissed and lied, been married and murdered, loved and lost, and for perhaps 500 years or so it has dutifully filtered their air, fed the soil and millions and millions of living things that have danced through its towering boughs, singing, buzzing, jumping and flying as we fought the French, the Germans, as we found the means to fly, drive, as we wrote, and painted, and danced, as we rock-and-rolled and went online, it has stood putting the world to rights, giving, giving and giving. I am in awe.

If my big tree is 500 years old, then it would have sprouted as Henry VIII was watching the jousting on the Field of the Cloth of Gold. It may be younger, let's say 350 years old, in which case it germinated as Charles II was signing a secret peace treaty to end hostilities with France and re-Catholicise England. How many beech saplings rose that year? How many just in the wood where I sit? An incalculable number, and yet just this one old tree remains. All its cohort have been eaten, felled or grown old, died and rotted. It's one in millions, maybe hundreds of millions; it's special, it's been nurtured by a sympathetic nature and spared by generations of deer, humans and fungus. Making ancient woodland is slow and chancy and that's why planting all those saplings in the projects that Megan has outlined above is no real guarantee that they will ever prosper.

In 2019 it was revealed that thousands of trees planted along the controversial HS2 route would have to be replaced after they were not watered in the previous summer's drought. Up to that point 350,000 saplings had been planted to compensate for the losses of ancient woodland this rail line is inflicting, but local farmers found that 80 per cent of the young trees on their land had died. HS2 Ltd said replacing the dead trees was more 'cost effective' than watering them. A total of seven million new trees, a mix of oak, hazel, dogwood and holly, are allegedly scheduled to be planted, but with this amount of aftercare it's obvious that the chances of any of them ever becoming woodland, let alone ancient woodland, are negligible. It's a statement of the bleeding obvious that you can't compensate for the destruction of ancient woodland by putting tiny saplings in the ground and abandoning them to their fate. This £106 billion project (from a January 2020 estimate) is set to destroy all or part of five internationally designated wildlife sites, 33 Sites of Special Scientific Interest, 21 local nature reserves, 693 local wildlife sites, four nature improvement areas, 22 living landscapes projects, eighteen Wildlife Trust Reserves and 108 ancient woodlands. These habitats will not be 'rebuilt by HS2 Ltd'; some could be by conservation agencies, who are at least the sort of people who water the trees they plant. And that's what I do, water the trees that I plant.

I live in a wood, so you may think it odd that I plant trees. Not many, last year about twenty, in recent times about a hundred. But guess what? At least 90 per cent of mine are still alive. I keep out the grey squirrels and hordes of deer, stop the grass from swamping them when they are tiny and give them a drink when they are thirsty. They get 'private tree care'; something we could all offer even if we don't have a garden. There are plenty of planting schemes, but the Woodland Trust will even give trees away for free to schools and communities. They also sell trees for under £10 (free postage) and provide great advice on planting and watering etc. That said, elsewhere you can get 'whips' for as little as £1.50 or, if you go for a walk in the late summer or autumn, seeds for free. At my 'Euston. We have a problem' anti-HS2 demonstration in September of 2019, I handed out hundreds of acorns to supporters and passers-by and told them how to germinate and pot them. This summer I've been sent photos of those baby oaks growing in gardens and in community spaces. That made us feel good.

The English oak supports more species of bacteria, fungi, lichens, free algae, mosses, vascular plants, invertebrates, birds and mammals than any other native species of tree. A study published in the journal *Biological Conservation* in 2019 entitled 'Protect Oak Ecosystems' revealed 2,300 species associated with the tree, though that figure did not include bacteria or microorganisms, meaning the real number will be much greater. Of these, 326 were completely dependent on oak and a further 229 highly reliant on the tree. A worrying 555 species are at risk from any decline in oak health and abundance; these include the oak lutestring moth, oak polypore fungi and oak leaf-roller beetle.

Our oaks need us, so grab some pots, collect some acorns, put them on a damp cloth in the fridge for a couple of weeks and then plant them 2.5 centimetres down in some rich soil. Love them tenderly for a year or two, re-potting as appropriate, and then find somewhere safe to 'release' them, away from mowing, flailing, cutting, nibbling or spraying. And then most importantly of all, name them and give them to a child. As they grow they'll compensate for a little of that child's inevitable carbon footprint and act as a personal point of reference and, if watched over, one that will last their lifetime.

HOW OLD IS OLD?

Do you remember the recent excitement when scientists used radiocarbon dating to determine the age of a Greenland shark? What they found made headlines around the world as the species took the title for the longest living vertebrate, with a female shark aged 393 years old, making her year of birth around 1620. Bowhead whales are second, with records detailing a 211-year-old individual. Their longevity is attributed to the fact that both species have slow metabolisms and live in very cold waters. Other than vertebrates, off the coast of Iceland a quahog clam was discovered that was 507 years old, and studies now indicate that some corals can survive up to 5,000 years! Unsurprisingly, corals are now believed to be the longest-living animals. It is very impressive, but the following new piece of research puts things into a whole new perspective ...

A research team at the Japan Agency for Marine-Earth Science and Technology had been collecting ancient sediment samples from the seabed of the South Pacific. This area is very nutrient-deprived, so you'd expect little life to be sustained over long periods of time. However, the researchers revived microbes – single-celled organisms – that had been lying dormant in the sediment that were over 100 million years old. These microbes were alive when dinosaurs roamed the Earth, yet once incubated in 2020 they were able to grow and multiply effectively. The lead author of the study, Yuki Morono, said, 'when I found them, I was first sceptical whether the findings are from some mistake or a failure in the experiment, but we now know that there is no age limit for [organisms in the] sub-seafloor biosphere.' It just goes to show that even the simplest of living organisms are anything but simple. Life can, and will, persist even in the most deprived environments.

Two romantic tree stories to fuel your arboricultural enthusiasm ...

I was once led by an eccentric local guide, dressed like Carmen Miranda and equally ebullient, to a small arboretum on a Caribbean island. It was scorching hot and her evanescent passion to satisfy my biological interests was admirable, but I was in the grip of lethargy when she cantered through the gate in the broken picket fence and into the ramshackle yard. With the last vestiges of waning grace I could muster, I accepted a jam jar of homemade lemonade from our host, a very wizened very old lady. Once temporarily refreshed she led us around a maze of freshly swept dirt paths, which weaved through her large and tangled wooded garden, pointing out flora using only unknown scientific names. She told us of their medicinal uses and snippets of folklore.

At the end of the trail, and close to the end of my body temperature's tether, stood a magnificent silk-cotton or kapok tree, one of the tallest and largest trees in the world, with some reaching more than seventy metres tall and three metres in diameter. The old lady leaned against the one of the characteristic buttress roots and gazed up into its towering canopy where the sun's cruelty was partially rebuffed. We all stood admiring this astonishing organism, listening to the deafening reel of the cicadas, until after a minute of reflection she reached into her pocket and withdrew a small framed photograph, which she handed to me. In faded black and white a sturdy man was stood in a parched field holding a sapling and opposite him a small girl crouched with a watering can – her and her father. I saw that tree twenty years ago, so the woman will have died by now, but silk-cotton trees can live between two and three hundred years. I hope someone continues to honour his and her investment.

In Sumatra, the last few members of the Orang Rimba tribe of hunter-gatherers try to hide in their shrinking rainforest home. I've twice been fortunate to spend time with them, albeit brief, but those encounters have profoundly impacted my life.

They were the most beautiful people I've ever met, living as they did in sustainable harmony with their environment. They had a habitat – think of that, humans with a defined habitat – into which they had evolved and coexisted with myriad other life. While walking on a muddy trail through the great emerald cathedral of trees one of the men stopped by a particularly notable and statuesque trunk and stretched out his hand to touch its bark. This, he told us, was his 'birth tree' – it was here that on the day he was born his family had buried his and his mother's placenta. The tree carried his name and everyone in their tribe knew and honoured it as they passed by. Later he stopped again, this time alongside an even larger specimen wrapped in a thick beard of moss and ferns. That tree was his father's 'death tree'; he was buried beneath its sprawling web of chunky roots. It too was sacred, a living tombstone where the man had become part of the ecosystem that had nurtured and sustained him. I think that if there were 'birth trees' and 'death trees' on the route of HS2 they probably wouldn't be so easily felled. Please name your tree.

As an aside, by its own admission HS2 won't be carbon-neutral until well into the next century; indeed its carbon footprint transgresses the government's commitment to the Paris Climate Agreement and its own carbon reduction targets. There will also be fewer and slower trains than first promised, and given how Covid-19 has impacted the way we do business, by Zoom rather than face-to-face, if and when it is running somewhen between 2028 and 2031 (January 2020, pre-recession estimates) many fewer people will be wishing to get from London to Birmingham minutes faster than before. Curiously the most significant projected time savings are from Manchester Airport, so yep, this rail project is also set to facilitate airport expansion. Surely on these grounds alone, though there are plenty more, this monster should be stopped in its tracks and rethought ...

While it's always all too easy to see the climate as something too big and too intangible for you and me to ever to be able to impact positively, I would suggest that very obviously isn't the case when by allowing projects like HS2 we are all too easily impacting it negatively. If collectively we can harm something as big as the environment, then collectively we can halt the harm or even heal something that big. It is not beyond our control, it's just that it seems to be beyond the realisation and ambition of too many people at this point of crisis. Something that scientists and campaigners are trying hard to rapidly rectify. I think the fundamental problem is that we grow to see the world as a very big place physically; too big to imagine the scale of in a meaningful and measured way. Think of Australia; that's on the other side of the world, the opposite side of the sphere. It takes bloody ages to fly there, up to a day in the air. And both the indigenous and naturalised people can appear so different from us; they sound a bit different, think a bit differently, they are 'foreigners'. In fact, the distance from Dover to Broome in Western Australia is 8,508.593 miles, while the distance from Land's End to John O'Groats is 601.477 miles, meaning that Australia is only 14.146 times the length of the UK away. Have you ever driven from London to the Lake District for a holiday? That's 262.828 miles and if you were lucky to avoid traffic (but God knows how) that would have been a reasonable journey. If you were to do it 32.373 times you could have theoretically driven to Australia. You can see where I'm going here – it's not that far to the other side of the planet, our Earth is not that big, so it's no surprise that while the people that live there may be politically and nationally 'foreign', biologically they are exactly the same species as me and you. They are our neighbours. Now imagine your neighbours from next door, in fact the rest of your neighbourhood, are utterly incinerated in a raging wildfire that burns for nine months, scorches an area more than eleven-and-a-half times the size of Wales, kills at least 34 people, destroys more than 2,779 homes, kills 1.25 billion wild animals and leaves livelihoods and entire ecosystems in ruins. Do you think you would be feeling a bit closer to the climate catastrophe? I think the reason we in the UK are not yet all on the streets with Extinction Rebellion (XR) is that too few of us are in real pain. But sadly not none of us . . .

Many parts of England were drenched with above average rainfall during October 2019, and in some places it was over double the average monthly total. This meant that the soils were wetter than usual for the time of year so that when, on 8 November, torrential rain fell on the saturated catchments across the southern Peak District, it caused flash floods and the rivers burst their banks. The next low-pressure system arrived a week later and brought more heavy rain to areas further to the south of England, raising water levels on the Rivers Severn and Avon. And so it went on until we experienced the wettest February in England and Wales since records began in 1766. Severe flooding occurred over many parts of the UK thanks to Storms Ciara and Dennis. By the time it abated there had been more than 5,600 flood warnings and thousands of homes were underwater, with some families still unable to return or re-open businesses by June. More than eleven people died, and the economic impact was said to be in the region of £1.8 billion. Insurers estimate that for every pound spent on defences, around nine pounds in property damages and wider impacts would be saved. And given that this is just the vanguard of our changing climate's wrath, we clearly need some action.

In the spring 2020 budget the chancellor promised to double capital spending on flooding, from £2.6 to £5.2 billion, which the government said would protect an additional 330,000 homes. If it happens that would be good news, but I'm afraid that only a small proportion of the money is likely to be spent on natural flood management projects that would protect wildlife as well as those homes. Wildlife and Countryside Link, which umbrellas a group of conservation charities, have said they estimate that only a pitiful 1 per cent of current spending on flooding filters down to nature-based solutions, such as restoring water meadows and wetlands, which act like a giant sponge to store water when flooded and as natural reservoirs during periods of drought. Those burned and barren grouse moors and the drained and 'sheep-wrecked' uplands could be rewilded, some of those as yet unplanted 30,000 hectares of government-backed reforestation could go towards rebuilding wet woodland and bog ecosystems, which also prevent the rain cascading down into the lowlands and into people's lounges and offices. And did anyone say beavers? Rather than prevaricate any longer why not airdrop these ecosystem engineers into river catchments all over the UK and let them alleviate flooding FOR FREE! (And when I say 'airdrop' please, *please* check out the parachuting beavers who were dropped into the Frank Church River of No Return

Wilderness, Idaho, in the 1940s – not that I'm suggesting we repeat such outrageous airborne invasions, of course.)

Over the past year we have come to terms with the tragic and inescapable fact that coronavirus will kill millions of us. It has become a tangible everyday threat to us all and we are justifiably frightened. Our changing climate is, for most in the UK, still not that tangible – we call it an 'existential' crisis, but the stark truth is that its threat is even more serious than this pandemic. We will develop a vaccine or a cure for that, we always have, and now with the global will, ingenuity and medical science and technology we have at our disposal it will not be that long coming. Not soon enough for too many of its victims, but in time to save humanity. But if or when our climate heats up beyond a certain point there will no longer be any doubt that billions of us will die – billions. So that should be even more frightening, but because it's going to happen 'tomorrow', and it's a more complex issue than just a single virus that we can name and identify, we just can't seem to feel enough of that fear to motivate us to act. Such a human trait; we always have to trip up before we mend the hole in the road. So good at cure, so poor at prevention.

IT'S WRITTEN IN THE SKIES

Across the UK on 21 June 2020, many people were astounded by the appearance of unusual blue shimmering wispy clouds in the night sky, just after sunset. There are many types of clouds in the atmosphere, but these were something quite different.

Noctilucent clouds, also called night-shining clouds, form in the middle of our atmosphere, or the mesosphere, where there is little moisture and very low temperatures. They are the highest clouds, existing 50 miles above the Earth's surface, on the edge of space, and form when water vapour freezes the smoke particles from meteors burning up, which then crystallise. They were first noted in Indonesia in 1885 when Krakatoa volcano erupted, releasing a surplus of atmospheric water vapour, and during the twentieth century, sightings of these extra-terrestrial clouds have been increasingly common and they appeared to be getting brighter, so scientists began questioning whether there was any correlation to climate change.

Using climate models and satellite observations, researchers simulated how greenhouse gases from the burning of fossil fuels may have contributed to the formation of noctilucent clouds over the last 150 years. The results indicated that since the late 1800s, methane emissions have increased concentrations of water vapour in the mesosphere by 40 per cent. By producing and releasing methane from human activities, these types of clouds are becoming more frequent. Franz-Josef Lübken, the atmospheric scientist working on the project, said, 'We speculate that the clouds have always been there, but the chance to see one was very, very poor, in historical times.' The conditions must be perfect to see them, as they only form in the northern hemisphere during the summer at mid to high latitudes, when mesospheric temperatures are low enough for the ice crystals to form from the water vapour. Additionally, you'll only spot them at dusk or dawn when they're illuminated by the sun under the horizon line. This is the first evidence that human actions are reflected by changes of clouds in the mesosphere. Lübken added, 'Whether thicker, more visible noctilucent clouds could influence Earth's climate themselves is the subject of future research.'

While we have been struggling to come to terms with coronavirus we've unsurprisingly taken our eyes off the climate catastrophe – not that we could have possibly missed its effects. I'm typing indoors – outside it's raining hard, and that hard rain is gonna fall all week somewhere in the UK. Storm warnings are in place for most of England, Wales and large parts of Scotland. We are in for more sticky heat, thunder, lightning and heavy downpours. Last week's temperatures reached above 34 degrees Celsius for six days in a row, the first time they've done that since I was born. Following last summer being the hottest on record for many European countries, heat records have been broken again and a prolonged stretch of sultry weather has given way to violent storms. The Arctic has experienced the highest temperatures it's ever recorded this summer, and here in the UK this year also looks set to be the hottest on record. The impacts of these extremes are destructive; this week there was a train derailment following storms and resultant landslips in southern Scotland – three people were killed and many more injured, an example of what experts claim will become more frequent events unless we take urgent action to make the UK more resilient to the impacts of the climate crisis. And there is something afoot ...

A new Climate and Ecological Emergency Bill has been drafted to demand the UK government act with emergency measures on the twin crises. It comes a year after the British parliament declared a Climate and Environment Emergency and, if passed, would substantially amend the Climate Change Act (2008), which is very obviously out of date, and necessarily accelerate the speed at which the UK has to act to address this issue, its most important concern. It has been written with contributions by highly respected climate, energy and ecology academics, as well as a leading author from the UN's Intergovernmental Panel on Climate Change (IPCC) report, and aims to hasten the government's action in regard to the climate and ecological catastrophe into law. So, what's in it?

Well, if passed the UK government would have to come up with a plan to significantly reduce its carbon emissions in real terms with instant effect (entirely necessary or we will meet none of our own or the Paris Accord's targets), then immediately calculate the carbon footprint of the entire UK, including those emissions caused internationally through import and export (which miraculously never appear in all the government's data). They'd also have to come up with a real, practical and measured plan to conserve and restore nature and wildlife in the UK with a keen focus on biodiversity loss (isn't that what we expect Natural England, Natural Resources Wales and Scottish Natural Heritage to be doing anyway?), and next, to reduce the threat to nature along all of the UK's economic supply chains both here and internationally (yes, we should actually consider our global consumption and its impact upon wildlife beyond our shores); also reduce a fanciful and unrealistic reliance on carbon capture technologies that haven't even been invented, to restore natural carbon sinks (like by actually planting those trees they've promised, and actively promoting and funding rewilding projects). Finally, they would have to form a Citizens Assembly, thus bringing social equity into decision-making, so that we ourselves can have a say in our future (just like XR have been demanding for over a year).

Supporters hope that the Bill will be introduced in the House of Commons when it reconvenes on 1 September, but there is one – shouldn't be an issue, but it's a huge one – problem. It's a Private Member's Bill and therefore requires one lead MP to table it (easy) and eleven sponsors (relatively easy) to be put before parliament, if they allow it time. But eventually this worthy, well drafted, highly informed and desperately urgent bill will require a majority in the Commons before being passed into law. Ah, now, ummm, yeah ...

We all know it should happen, particularly after this June's chilling report by the government's Climate Change Committee, which revealed with horrific candour that the UK's net zero by 2050 target gives only a 'greater than 50 per cent' chance of averting climate catastrophe, and recommends preparing for a two to four degree Celsius increase in global temperatures, for which read ... Armageddon. Such an increase would cause massive food and water shortages, irreversible loss of biodiversity, the complete destruction of ecosystems and the complete inundation of coastal cities. For which read ... the end of life as we know it.

This Bill should also be important because it addresses another now-glaring inadequacy in the 2008 Climate Change Act: that it has no regard for ecological collapse. It's now so very obvious that both the climate catastrophe and the mass eradication of biodiversity are intrinsically linked. We can't save species or ecosystems if we can't stop the fire, we can't have polar bears with no ice, and without ice we can't cool the planet.

The Bill is being supported by thinking, intelligent politicians and the NGOs and XR, but I fear that we – yes that's you and me again – will have to make some noise about this brave but essential piece of prospective legislation. As I've said elsewhere, in these exceptional times we have an unfortunately unexceptional crop of global governors – so dig in, it's not a case of your country needing you, our world needs you now.

FINDING HIGHER GROUND

Much of our attention has rightly focused on the broad aspects of the climate crisis and how it will reshape and alter the planet – temperature and precipitation changes and species population fluctuations. But we also need to be vigilant about the specifics of biodiversity redistribution. Research into the movement of wildlife driven by changing conditions is quite patchy, but one thing we know for certain is that ranges will shift north and upwards in altitude as typical home ranges get warmer. It can be very difficult to motivate local actions to support wildlife on the move when the vast majority of research looks at global redistribution. To address this, and to learn which species and habitats to prioritise locally in the UK, scientists at the University of York used social media, among other methods, to gather citizen science data about which unexpected species were turning up where.

The study, conducted between 2008 and 2018, found that at least 55 species in the UK had already shifted their ranges due to climate change; this included the southern emerald damselfly, the green-jawed tube web spider and the purple heron, who all expanded north. However, this number is most likely a big underestimation and the study suggests we prepare for dramatic changes in species composition. This shift could cause disruption and additional ecological pressures to the new areas that species are moving to. The paper found that 29 per cent of species with altered ranges were having negative impacts, like increased disease transmission and outcompeting resident species, but 20 per cent were having positive impacts through ecotourism. Currently, they suggest that the negatives and positives are balancing one another out, but that won't always be the case as more animals start to shift.

Studies like this with the use of modern techniques will really help us to understand exactly how species are moving and how we can better shape conservation strategies and programmes to accommodate all – indeed, ten of the 55 species were identified in the study because of the use of hashtags on Twitter, which goes to show how useful social media may become in monitoring populations and environmental trends. The nature that we are accustomed to seeing in our back gardens or parks, or even in the country, is going to change, and the more we know specifically about that change, the more support we can provide in our green spaces.

But of course it's not all about us. We are not the be-all and end-all of life on Earth; we are an organism that is intrinsically entwined with, and therefore dependent on, our planet's ecology, one that would be secure and sustainable if it weren't for our excessive abuses. It's said we are precipitating the sixth mass extinction. I wouldn't argue that's not the case, but I am again uncomfortable with the lazy vernacular. Just as we haven't 'lost' species or their habitats, we've destroyed them, it's not an extinction event we are fuelling; it's a calculated and organised extermination. And while we are on words, we have come up with 'ecocide' to describe the mass killing of animals, plants and fungi, of forests, rivers and seas, etc, etc, essentially the killing of ecosystems. But I think we are doing more – we are killing everything. We are committing 'omnicide'.

Now, etymologists may complain that the 'cide' component implies a murder or a killing, and what we are concerned about here is a catastrophic natural event, but given that we are culpable of executing it, and at this point with conscious intent, I'm happy with 'cide' and unhappy but resigned to recognise contemporary *Homo sapiens* as 'omnicidal maniacs'. Not all of us though – some, indeed more and more of us, are trying to halt and heal the victims of this crime. But who are the criminals? Who is killing Planet Earth?

I don't think we are living in a Bond movie; I don't think there is a person or a gang of people actively plotting this extermination event or gloating over their successes to date, satisfied by the destruction of 60 per cent of the world's wildlife in the last 50 years. No, this omnicide is a complex and layered case, one that leads to a very uncomfortable conclusion. The plot is thick, and sick ...

Let's start with all the usual suspects: Trump, Bolsonaro, Morrison, Putin, Johnson – all those senior elected representatives who repeatedly refuse to listen to scientific advice and fail to effectively warn us of the true gravity of the Climate and Environment Emergency, which at least some have signed up to. The same cadre of global capitalists who refuse to mitigate against the apocalypse by actively encouraging or turning a blind eye to everything from dirty energy generation, promoting an ongoing dependence on fossil fuels, overfishing, ocean pollution, intensive farming, rainforest destruction, all the way up to unchecked human population growth. Their stupidity or short-term greed is killing Planet Earth. Guilty.

But to get away with their crimes they need to keep us in the dark, confused or in conflict, and thus their erstwhile allies are those media tycoons who dispense fake news and a relentless denial of the crisis. Through their organs of mass manipulation, TV, newspapers, the internet and now social media, they ignite a fear of change, feed a culture of ignorance, nourish divisions between us and distract us with relentless trivia. Turn on the news, you'll find more about celebrities' bust sizes than bushfires – real news is being driven underground, it's becoming counter-culture, not everyday fare. Guilty.

Money makes their world go round, so the politicians rigorously guard the mega-corps, however destructive their industries. They prop up toxic chemical manufacturers, the fossil fuel industry, those media moguls, and all those financiers who continue to invest in these monsters and repay the politicians themselves. They have also built an almost impenetrable legal system to make laws to protect this ghastly machine, which gyrates around the need to maximise short-term profit at any cost, when its true cost is to the environment because to keep it spinning requires constant growth in consumption. Guilty.

And now as I prepare to pull off the mask and reveal the true culprits, we can be momentarily distracted by those people and cultures who consider themselves dominant over, and at liberty to exploit, animals and the natural world so that we can continue to prosper because everything else that lives is just here for us to use and abuse. Guilty.

But as the planet police swoop to arrest those who are killing the world there's a last-minute twist; they are not breaking down the doors of the White House, or Shell or Exxon's headquarters, or storming the offices of Bayer, or Fox News, or the Industrial and Commercial Bank Of China Ltd – no: they are walking up your driveway. You see, we are the consumers who, albeit often unthinkingly, prioritise our selfish interests over the health of the planet. We are the ones who collectively lack the courage or moral strength to make the social and economic changes that would ensure that our children have a world fit to live in. We are the ones who are so easily distracted that we can't be bothered to make independent or well-informed opinions, who would rather comply with the system than risk any uncomfortable conflict. We are the ones content with our cosy lifestyles, too scared to change. Guilty. We are guilty of killing Planet Earth; we are the omnicidal maniacs and we will be held accountable.

Of course that 'we' is not necessarily you or me. Phew! Or at least, it's not all of you and me, but it is some parts of us. Because as much as we might try to be as ecologically and environmentally responsible as we can, we are still all consuming far more than is sustainable. If all the 7.4 billion people on the planet were consuming at the same rate as we do in the UK, we would currently need two extra Earths' worth of resources to keep going. If we were Americans, it would be four extra planets. We can feed our birds, switch to greener energy suppliers and bike to work, but unless we cut that consumption we are all in trouble. And that consumption has a global basis – we are not eating the produce and gathering the fuel from our local communities, it's coming from all over the world, grown or harvested from a range of ecosystems we can hardly even imagine. So as much as we may put enormous energies into conserving our own parts of the environment, and can do a remarkable job of protecting and restoring them, we are perversely, on many occasions, exacting a more direct impact on other not-so-very far-flung parts of the world. And it's this connectivity that we should all be more conscious of. As you down tools having done a good morning's volunteering on your local nature reserve, sit down beneath a magnificent old oak, covered in moss and lichens and ferns, full of invertebrates and the birds that feast on them, and open your lunchbox and nibble a biscuit, you are about to become part of another forest that was. The palm oil in that snack has come from Indonesia, from a plantation that, until very recently, was tropical rainforest and home to orangutans, tigers and many times the diversity of insects that buzz around your oak.

This is one example, but there are many and, like me, once aware you may have tried 'economic activism', by stopping the purchasing of those products with detectable or direct links to environmental harm. Avoiding pharmed salmon is easy, avoiding palm oil is not. In the wake of a programme I made for the BBC in 2018, *In Search of the Lost Girl*, which, along with their excellent

Red Ape documentary, threw an uncompromising spotlight on the massive destruction the oil palm industry has wrought and is wreaking on the East Asian tropical forests, I sought to expunge it from my life. And failed. To say 'it's in everything' is only just an exaggeration and the appalling, but undoubtedly intentionally poor, quality of food labelling makes any deliberation over buy-or-not-to-buy impossible, certainly when it comes to ingredients. Some food labels lie, some are conservative with the truth, few present boldly what we need to know to make easy and instant choices in those aisles of complacent temptation. I'd like to see 'THIS CONTAINS PALM OIL', or, 'THIS CONTAINS GLYPHOSATE', or, 'THIS HAS BEEN UNSUSTAINABLY HARVESTED', or, 'THIS CONTAINS TRACES OF A COCKTAIL OF CHEMICALS THAT HAVE SEVERELY DAMAGED AN ECOSYSTEM' printed in bold red type alongside the price and another figure that defines the real environmental cost in pounds and pence. Only then could we consumers use the power of the pound in our pocket to maximal effect to punish the people whose businesses are killing our planet.

A moment of honesty ... Every night when Sid and Nancy and I go to bed I give them a treat. It's a guilt thing – I'm typing now, not giving them the attention they crave and deserve as the dogs that I love. The treat is a vegan biscuit, but it contains palm oil. Every night I revel in their relishing part of some distant, dark, dripping forest that was felled, burned, drained and densely planted with non-native palms; every bedtime we three in our remote New Forest cottage expedite the impending extinction of one of the planet's most beautiful creatures. I switch off the light and dream of orangutans. What a nightmare.

I can't live with that. And nor can our children.

Honey, the kids are rising

There is undoubtedly a growing frustration and anger among us all – or at least those with an environmental conscience – when we are faced with daily injustices, crimes and declines. It's a constant stream of bad news: yesterday, a white-tailed eagle was found poisoned, sharing the fate of many before it; today, 4,000 tonnes of oil spilled into the waters around Mauritius, damaging coral reefs; and tomorrow ... well, possibly an unlivable climate with soaring temperatures, floods, droughts and food shortages. It sounds pretty scary, doesn't it?

It sometimes feels like the issues are too big, too complicated and too uncomfortable to discuss and resolve. We are so afraid of potentially upsetting anyone that we are crippled by our own thoughts, rules and regulations – unable to make positive progress. Many of us are fed up, irritated and overwhelmed. I know I am. Eco-anxiety is a term used by mental health professionals to describe a 'chronic fear of environmental doom'. It is not yet listed as a diagnostic disorder, but research shows it can manifest in many ways, such as trauma, PTSD, depression, substance abuse, aggression and the overwhelming feeling of helplessness and fatalism. In all honesty, for me I feel helpless and pretty depressed at times, but some days are better than others. It's important that we discuss this openly with one another and remember that, if needed, there's a growing number of psychologists who are getting trained to discuss and support those suffering with such anxiety around environmental and climatic concerns. We can't change what happened yesterday but we can learn from it. Tomorrow depends on how we behave today, and if we are capable of making some drastic lifestyle and systematic changes, then we have the solutions and ability to change our trajectory. We've just got to wake up, get up, get out and get on with it.

In 2018, a new movement arose that turned that frustration and anger into something productive and powerful. It's a movement formed by young people – generation Z – who have grown up in a world where exposure to the impacts of the climate crisis have been unavoidable, leaving many uncertain about the shapes of their futures. Eighty per cent of teenagers today say they feel under pressure to save the planet, but aren't sure of the best way to go about doing so. But one individual didn't let that stop her because doing something, anything, was better than doing nothing.

On 20 August 2018, aged fifteen, Greta Thunberg skipped school to sit alone outside the Swedish parliament to highlight the gross political inaction on climate change. Just six days later, she was joined by a few other students, some parents and even a few teachers, who sat by her side in solidarity. The media quickly caught on as mainstream news around the world and social media was flooded with images of a young girl with long braids in a yellow coat clutching what soon became the iconic placard reading 'SKOLSTREJK FÖR KLIMATET.' I don't know Greta, but I doubt she could ever have imagined what came next. Today, just over two years since her first demonstration, fourteen million people inspired by her have skipped school on Friday mornings and hit the streets in protest. They have come together from every single continent (yes – even Antarctica) in 214 countries, from Abkhazia to Zimbabwe, in 7,500 cities for over 80,000 peaceful school strikes. The movement is aptly named Fridays For Future.

Greta has since sailed across the Atlantic to address the UN

Climate Action Summit, spoken at the UN Climate Change Conference and addressed world leaders. She has also received many honours, including being listed by The Times *as one of the Top 100 Most Influential People, been included in the Forbes list of the World's 100 Most Powerful Women and has been nominated for the Nobel Peace Prize two years in a row. Most recently she accepted $100,000 (£80,000) as an award from Human Aid, which she immediately donated to the Coronavirus Relief Fund. Personally, I am in awe of anyone who can look our leaders in the eye and say 'how dare you' with such conviction and passion. There's no ulterior motive. It's just about science and genuine concern for our future.*

Fridays For Future is formed by millions of young activists who all share that heavy burden. Yet they have harnessed that negative energy and turned it into something productive and positive; all because of Greta, who taught us that one person does have the power to change the world. Their demands are simple:

1. *prevent the global temperature from rising above 1.5 degrees Celsius compared to pre-industrial levels*
2. *ensure climate justice and equality*
3. *listen to the best united science currently available.*

All the things you'd expect our governments to be prioritising anyway, but no, young people have to miss school to remind them of what's important. It's a movement that has created an international awakening and it's rippled through all generations. I have joined many of these demonstrations, but I vividly remember my first experience of one in London in 2019, where over 100,000 protesters of all ages marched to Parliament Square.

Everyone was connected by a common cause and it was genuinely emotional to witness. It was a festival with a mission that showcased creativity, music, art, science, compassion and reason. I had to take a step back to take it all in – it felt like the beginning of something very important. And I think it was.

No one in the UK has been more committed to Fridays For Future than Holly Gillibrand. Holly is fifteen years old and lives in Fort William, Scotland, and has been missing one hour of school every Friday morning since the beginning of 2019. To date, she has striked for over 80 weeks in a row during the snow, rain or shine (and even from her house while isolating during Covid-19). Holly told the BBC that 'the sacrifice is a small price to pay for standing up for our planet. If you get detention, that's nothing to how we will suffer in the future if nothing is done. I want to get Scottish leaders to take climate change seriously and know that they're destroying my future.' Her profile has grown extensively through the use of social media to connect with other like-minded people and in 2019 she was named as 'Scotswoman of the Year' by the Glasgow Times. *Holly received the award on the same day that the Scottish government released their climate emission 2018 statistics, which detailed how Scotland had failed to hit their target as emissions increased by 1.5 per cent between 2017 and 2018. Holly said at the time, 'we are heading in the completely wrong direction'. But that did not stop her from fighting for what she believes in. Today Holly is a huge advocate and spokesperson for multiple rewilding protects, such as Heal Rewilding and SCOTLAND: The Big Picture, and shares her message as a columnist for the* Lochaber Times.

We have many young naturalists and campaigners in the UK who are leading the way for positive change, with some of whom

I hope you are familiar – Dara McAnulty, Finlay Pringle, Anita Okunde, Bella Lack, Anna Kernahan, just to name a few. It's their futures that are most at risk. It's sad that we turn a blind eye (perhaps because we feel guilty) and those in positions of power are more interested in profit over planet. My favourite placards from the Fridays For Future movement read, 'I'd be in school if the world was cool' and 'Why should we go to school when they don't listen to the educated anyway?' Well, now it's time for us to open our eyes and either join them, or take a step back, because either way the kids are stepping up.

NECESSARY TROUBLE

The first rally I attended was the brilliant Rock Against Racism march in April 1978. I wasn't alone; 100,000 people who were rightly sickened by the National Front ended up in Victoria Park where we were thrilled by The Clash, X-Ray Spex and the inimitable Jimmy Pursey. I was joined in the crowd by Billy Bragg, and plenty of other angry and creative young people. Being in that punky throng doing what I could to fight racism energised me; I saw the power people could muster to effect change and it made me realise that I must start to fight for other causes that I cared about, principally the destruction of the natural world.

I grew up in a suburb of Southampton, and in the 1960s and on into the 1970s there were lots of patches of what was known as 'wasteland' scattered around between the houses. Most were on land that was less suitable for building, such as recently defunct brickworks, steep sandy slopes and bombsites. None were wasted on me – these tatty, scrubby handkerchiefs of bramble and grass were where I cut my teeth as a naturalist, catching lizards, finding birds' nests and watching fox cubs. And obviously as I got to know them and their wild residents I grew to respect, value and love them. I developed a sense of connection; not so much ownership, but a sort of trusteeship. I cleared them of litter and tried to nurture and protect them.

Then came the eighties – the greedy eighties – and with them came the greedy housing developers into my backyards. They call it 'infilling', I'd call it stripping invaluable wild spaces and wildlife out of the lives and reach of the communities already dwelling there. So the letters went up on the lamp posts and the brambles got bulldozed and I started to hurt as the homes of all the species I knew went under the hammer blow of 'progress'. As they were successively destroyed I began to get angry, angrier all the time, until they came for my favourite reserve.

It was a sheer-sided bowl where sand had been scooped out to make bricks during my father's youth. I remember it being lined with tall elms before the Dutch disease came for them. They fell, their bark peeled and I scavenged for invertebrates there to feed my bedroom-crammed reptile menagerie, and one day I found a pit full of poo – my first badger latrine! Oh, the excitement! Only paralleled when, after a lot of long waits in the glow of the streetlights, I glimpsed the low-slung and grizzled form of the animal itself. I got to know those badgers, I fed them and studied them and snuck between the bracken at the backs of people's houses to monitor their sett. They had three cubs the spring that the council sold it to a developer. I knew they were a protected species and I knew that destroying their sett was illegal so, sat on the edge of my bed, I handwrote a letter to the council and to the builders. Both denied there were badgers at the site. I wrote again and this time they arranged to meet me on the pavement. I'll confess I was anxious as I stood there waiting. I was extremely socially clumsy, especially with strangers, and found it hard not to get overly aggressive and then completely lose my temper when any conflict arose. I'd arrived early and I'll confess the thought of just slinking off before they arrived entered my head several times. After what seemed an age – they were late – two men arrived. They had very shiny shoes on, not 'brambly badgery' footwear. I remember those four shoes very well; I looked at them throughout the course of our conversation, chit-chat that I naively thought was polite. I later realised it had been patronising. Anyway, I led them to the sett, they got dirty and snagged and scratched, and then summarily pronounced it as 'unoccupied'. I became indignant, I told them they were ignorant of the appropriate knowledge to make that judgement, they left and I spent days in a rage. Then a letter arrived ... 'no sign of life at what might have been a badger den' ... den! Pillocks!

My life changed forever when I opened and read that letter. I realised that I had no choice but to fight. I was 21 and had no idea that I'd spend the rest of my life combatting ignorance, intolerance, greed and liars in shiny shoes, but that piece of paper was the spark that lit up my determination to never back down when I was right and to never shy away from a battle to protect wildlife however big it might be. I went to the local press, they printed my photographs of the badgers, people wrote letters, despite my shyness I went door-to-door with a petition that hardly anyone wanted to sign and, eventually, and all too easily, the shiny-shoed people got permission to 'evict' the badgers. This they attempted to achieve by pouring diesel into their sett and then blocking the holes so they wouldn't re-enter after they'd left. I couldn't prove it, but I know that when those big yellow Caterpillar trucks crept over that mound those badgers and their three cubs were crushed and buried alive. So I lost, the badgers died, the houses went up and as if in spite they called it 'Badgers Rise'. But it was the buildings that rose, ironically, only to fall down again. Yes, the piling they drove into those sandy cliffs wasn't sufficiently substantial and they began to subside. Inconvenient for the purchasers, expensive for the developer. What a shame.

This experience hurt; I'm a very bad loser. I blamed myself, lying in bed running and re-running over scenarios where I could have been more effective. I was angry, but not only with myself, with all those so-called conservationists who just seemed to be standing by and letting this sort of thing happen. The badgers had been protected by law, but who was there to implement the law? Surely those who had written it should be ready to put it into action. That was our government, or at least those agencies which it had formed to protect our wildlife. So what of those statutory bodies and what of those politicians who we elect? Are they really doing what they should for our wildlife and the environment?

Personally I expect our elected representatives at every level, from the parish council to the parliamentary cabinet, to not only have a mandatory need to understand the gravity of those threats to nature and the environment, but also to be pre occupied with addressing them, particularly now they have declared a Climate and Environment Emergency. To not be able to do so, or to not do so, is no longer something we should countenance because clearly we are in deep shit and we have given them a mandate to govern, and an expectation that they do so responsibly, and in our interests. Now, being reasonable, it is unfair to expect all these people to be climate scientists or ecologists; it would be great if some of them were, but we are not at that point – yet. However, in the absence of expertise they should employ climate scientists and ecologists to inform them – and take that advice very seriously. That's how reasonable and rational people move efficiently through life – we consult experts. My oven breaks down, do I as a zoologist try to fix it? No, I consult an electrician whose knowledge and experience I value and then pay for. So instead of me bodging it and it burning my house down, it cooks my dinner. I'm honest about what I don't know, I enjoy finding people who do know; as Clint Eastwood told me way back in 1973, in order to effect magnum force 'a man's gotta know his limitations'. And, sadly, what we currently have in these exceptional times is a wholly unexceptional crop of politicians who clearly don't feel brave enough to acknowledge their limitations, and/or are just plain bad and only serving their own short-term interests. Before we discuss what we – that's me and you – can do about sorting out the planetary crisis, let's have a brief look at the UK government departments that ought to be helping us protect our wildlife and its habitats.

Firstly, due to devolved government it is impossible to make direct comparisons between the various UK statutory agencies responsible for wildlife conservation. They all have different roles and remits, and different structures. For instance, Forestry England, formerly the Forestry Commission, is separate from Natural England, but Natural Resources Wales has absorbed their forestry commission, as they have the Environment

Agency. Nevertheless, there are parallels in their problems – most notably those of significant under-funding and loss of trust from those of us who want to and are trying hard to save the vestiges of the things we love.

When I wrote my parts of the *People's Manifesto for Wildlife* in 2018, I said that Natural England were 'not fit for purpose'. Writing those four words caused me sadness. An employee of the Nature Conservancy Council, as he was between 1973 and 1991, Colin Tubbs was both a hero and then a mentor. I wouldn't be writing this book without his very constructive and formative input to my early life. From the age of fourteen to 22 my dream job was to work for that agency and try to emulate his studies in the New Forest, a place he taught me to love through science. But I'm afraid that in the interim this once-effective independent advisory body has not only been rendered impotent, but increasingly frequently also presents a significant handicap to conservation in England.

Since I wrote those painful words, Tony Juniper, a highly qualified conservationist, has attempted to take the helm of the ship, which I described then as 'rudderless'. Like many, I like Tony very much, and his credentials are sound (his books are excellent too), but my perception is that his admirable intentions have yet to deliver any dividends. You see, the board he presides over, but cannot ultimately choose the composition of, includes members with interests that potentially conflict with conservation of the natural environment. The body is also, unfortunately but necessarily, beleaguered by an enormous litany of Freedom of Information Requests and Judicial Reviews, despite the fact that we the public pay for its existence. And sometimes that information isn't forthcoming so it's fair to ask why we are having to sometimes fight through the Information Commissioner's Office for what should be our information. And clearly and frequently there is a desire to hide that information because it really doesn't fit with what should be an independent advisory body acting independently on behalf of the UK's wildlife. But then Natural England is sadly no longer that.

The once-considerable expertise of their staff has been undermined by these circumstances as over time they have been stripped of their ability to make informed decisions. Therefore many of NE's actions, or inactions, are embarrassing, inexplicable or, in some cases, even dangerous to wildlife conservation. It has struck deals with developers, grouse-moor owners and others with economic interests that conflict with nature conservation values, freeing them from regulatory restraint without any or sufficient ecological benefit in return. Monitoring of those jewels in our conservation crown, SSSIs, has been all but abandoned, and its wonderful network of National Nature Reserves is imperilled. Frankly it's terrible. If I had a magic wand and a deep pocket I'd wave it over Tony's compromised board and his bank account and do the decent thing. But that's not the job of an imaginary philanthropist, that's the job of a responsible government.

So that's England; what about Scotland, the home of so many of our natural treasures? Sadly (I know, that word is getting plenty of use) it's a similarly grim scenario – maybe worse. An ongoing and embarrassing catalogue of conflicts between Scottish National Heritage (SNH) and UK Conservation's obvious objectives dogs this body. The fiasco surrounding the Strathbraan raven cull in 2018 – which SNH sanctioned and which its own investigation described as 'completely inadequate' in a damning report into its validity – highlighted very bad decision-making. This has been usurped by another horrible fiasco surrounding the protection of, and then licensed killing of, the beavers on the Tay. It's got to the point that unless it can be explained as wholesale incompetence, there must be something else going on. Thankfully, the report into the now wholly abandoned raven cull plumped for the former explanation. But the 'major flaws' discovered in this review appear to extend throughout this agency and its practices, and many believe that SNH 'should be completely re-designed rather than (modified).' Despite all the desire and evidence to the contrary it has steadfastly refused to properly promote the reintroduction of the beaver nationally, continues to fail to protect those beavers on the Tay, which continue to be inhumanely shot and burned, has done precious little to address the on-going excesses of mountain hare killing on grouse moors, failed to engage in the issue of raptor persecution and, akin to its English counterpart, has undeniably and ineffectually presided over a continual decline in the wildlife under its jurisdiction.

When it comes to the situation in Wales I'm no expert – like I said, 'a man's gotta know his limitations' – so when I compiled the aforementioned manifesto in 2018 I asked my colleague Iolo Williams for comment. Here's what he said:

'A recent internal survey showed that only 14 per cent of Natural Resources Wales (NRW) staff are happy with the way they are managed. The Wales Audit Office recently queried

NRW accounts for the third year in a row. The chairwoman, Diane McCrea, resigned in July following the scandal of underselling timber to a single private buyer. NRW have constantly blocked attempts to reintroduce beavers to Wales despite the full support of all the major conservation organisations. Morale is rock-bottom with conservation staff leaving en masse and not being replaced. There have been dozens of major slurry pollution incidents on once-famous salmon and sea trout rivers in West Wales over the past twelve months, killing tens of thousands of fish. There have been NO prosecutions by NRW relating to any of these incidents.' He continued:

'Fundamentally, NRW needs individuals in the senior management team and on its board that are committed to our environment and its wildlife. At present, there is no respected conservationist in senior management. This would help tackle its woeful record on nature conservation and help solve its staff dissatisfaction difficulties. It also needs to overhaul Glastir, its completely ineffective agri-environment scheme. At present, its success is measured in terms of percentage of land in the scheme, as opposed to measured increases in target species. The prescriptions and monitoring are woeful.'

Pretty damning, but that was then. What about now? I asked Iolo for an update. His opinion was that not much had changed:

'They do now have a recognised scientist/conservationist, Prof Steve Ormerod, and an excellent young, conservation-minded farmer, Geraint Davies, on their board, but little has changed in terms of employing conservation-minded senior staff. There are ongoing serious agricultural pollution incidents that go unpunished, much to the disgust of local and national fishing organisations, but there doesn't appear to be any drive from NRW to seek a prosecution and a fine worthy of the environmental damage caused. So to sum it up, no real positive change.'

Lastly, but not leastly, because it has some of the most beautiful natural spaces I've ever explored in the UK, we come to Northern Ireland and again I defer to expertise. This time that of the ruthlessly truthful young naturalist Dara McAnulty ...

'Conservation in Northern Ireland is a story of great potential slowly being thrown away. There is an incredible wealth of diverse habitats here from mountain to sea, bogs, fens, wetlands, coastal dunes, pearl mussel and salmon rivers, all under threat from the usual enemies of development: intensive agriculture, pollution, lack of political will and public awareness. Most of Northern Ireland, including our designated sites and other priority habitats, are receiving levels of agricultural nitrogen that are significantly above their "critical load", the concentration at which significant ecological damage occurs. Northern Ireland is responsible for 12 per cent of total UK ammonia emissions, despite only having 3 per cent of the UK population and 6 per cent of the land area. Agriculture is dominated by livestock and cattle are responsible for 70 per cent of ammonia emissions.

In 2019, I helped launch the State of Nature Report with the RSPB. This outlined some horrific figures – for instance, of 2,450 red-listed species for the Island of Ireland, 11 per cent are at risk of extinction in my lifetime. Now that UK has left the European Union, it also remains to be seen how and if the newly formed Department for Agriculture, Environment and Rural Affairs (DAERA) will uphold the programme of measures it has agreed to implement under the European Marine Strategy Framework. At 650 kilometres the NI coast is relatively short, but supports an exceptional diversity of marine wildlife and habitats. I am particularly concerned about Strangford Lough, an important winter migration destination for waterfowl including 75 per cent of the world population of pale bellied Brent geese. The ban on scallop dredging came too late to save the sites' horse mussel beds, many of which haven't recovered since the dredging was finally stopped in 2003. However, large areas of sea grass beds and other important habitats remain. DAERA's management plans for Strangford Lough need to be implemented to preserve and restore one of the jewels in the crown of the UK's wildlife.

While DAERA is currently steering away from the disastrous "Going for Growth" strategy and towards a more sustainable "Green Growth", one wonders if it is still in the grip of old thinking. The current Environment Minister, Edwin Poots, has been signalling support for greener policies, but I am worried that he may still unleash a badger culling programme similar to the devastating culls in England. I am also worried that DAERA agri-schemes to create "Forests of the Future" and "Environmental Farming Schemes" may end up being just more monotonous conifer plantation and too little too late for our priority habitats. As Northern Ireland is the least-forested part of the UK, it would be a missed opportunity to revive our former great native woodlands.'

Ultimately the most tragic aspect of these agencies' declines is the escalating loss of trust between them and the wider conservation movement. This continues to grow and gain widespread attention; Natural England extending licences to disturb bats two weeks beyond that which is normal and into the height of their maternity/birthing period to allow HS2 to wreak havoc in our ancient woodlands, and yet more licences to facilitate the expansion of the badger cull, mean that few within the conservation sector now believe that they, or Scottish National Heritage and Natural Resources Wales, are properly independent or impartial. We all fear the commendable staff who remain have lost their voices, when they should be able to publicly speak their minds to governments. And I should add that the comments above are not directed at those highly capable, hard-working and committed employees of these agencies who, despite difficult circumstances, strive to be effective conservationists. They all have our utmost respect and admiration. But the situation is desperate, so what can be done to fix these agencies?

Well firstly, can or should we fix these agencies? As long as they are funded by neglectful governments we must ask, can they be secure and truly independent? Given the current predicament I'd say no. We cannot trust either to be acting in our best conservation interests. So, in the longer term – and longer-term thinking is what we are all aspiring to, isn't it? – I believe that such bodies need (and get this because it's important) very significant, ring-fenced, apolitically influenced long-term public funding. Idealistic or essential, you decide. In the short term there is absolutely no doubt that a major injection of public money, a complete re-structuring of their boards or councils to include properly qualified and independent ecologists, investment in staff training and retention, and complete transparency and access to data would perhaps reinstate some impartial influence and re-engender some respect in these bodies.

In regard to Natural England therefore, we'd be asking for nothing that the House of Lords didn't already request in March 2018 and that their departing chair didn't flag up in no uncertain terms when he retired in November of the same year. Former property developer Andrew Sells, perhaps a more curious appointment than Tony Juniper, highlighted that he'd lacked a human resources director, a finance chief and a communications department. Merging of its offices and IT facilities had spiralled following 'cut after cut after cut' and the underlying basis of his complaints was that NE was being absorbed by DEFRA, which then supplied nearly all of NE's funding. We all wished he'd said this when he'd arrived rather than on his departure, but what was clear was that his agency had been dismembered and disempowered and its staff were disenfranchised. During the Brexit preparations Michael Gove's department 'poached' or 'recruited' more than 2,000 staff, including 400 seconded or loaned from the Environment Agency and Natural England. At the time a DEFRA spokesperson said, 'The work of Natural England and its staff to protect our invaluable natural spaces, wildlife and environment is vital and its independence as an advisor is essential to this. With the government's 25-year plan for the environment, Natural England will continue to have a central role in protecting and enhancing our environment for future generations.' I don't believe them, do you?

So what do we do? Sit back, maybe moan a bit, post a disgruntled tweet? Or politely, democratically and peacefully take these very important matters into our own hands?

I think it's time to ask yourself some important questions. Such as, do you think that by feeding the birds in your garden you will be fully addressing the declines in the UK's woodland birds, which have decreased by 30 per cent between 1970 and 2018, and 5 per cent over the recent short-term period? By not mowing part of your lawn do you think that you will help to redress the destruction of our flower-rich meadows, 97 per cent of which have been ploughed up since the 1930s? And by cutting a hole in your fence to allow hedgehogs access to and from your neighbours' garden do you think you will secure a recovery for the species following its own 97 per cent decline?

You're going to have to say 'no', which I would say is true, but not wholly true. More of that in a moment, but now think for a minute and then ask yourself this question: am I an activist? Or more pertinently, can I afford not to be an activist?

I have a very strong feeling that you might be sort of an activist already. Because I think that there is a very good chance that you do feed the birds in your garden, that you probably have said 'no to the mow' and that you have cut a hole in your fence, or would if you were lucky enough to have hedgehogs snuffling through your undergrowth. This means you have actively done something to directly help wildlife. And that is undeniably brilliant and I'm sure it makes you feel brilliant too. And of course, because you won't be the only person to have done these things, then collectively we *are* making a difference – after all by offering more than £360 million of garden bird food every year we have all led to an increase in the bill length of great tits. And we've stabilised the decline in hedgehogs in some areas. But is all this enough to stop our countryside and its wildlife from going to hell in a handcart? No, I'm sorry, it's not. Because all the meaningful and productive things we do on our collective patches are still not on a scale that will stop the colossal rot.

But don't worry, because we have an arsenal bristling with ideas, techniques and abilities to stop the rot pretty much in its tracks; things like the rewilding and reintroduction projects we've written about here. But to win, to effect real change we need to reform the way our national parks are run, stop eating 'pharmed' salmon, reform UK shooting, ban certain pesticides, ban lead shot, support our farmers, stop the badger cull, and, obviously, stop fox-hunting too. You can't do these in your garden on your own, and sadly as you will have noticed many of the people we would expect to be trying to make these things happen are not, because they've lost their zeal for campaigning, if they ever really had it. So we cannot rely upon them, and in fact we are in so much trouble I think it's unwise to rely on anyone. I think the only way to guarantee to get things done is to do it ourselves – so that's me, and that's you. And to achieve what we need in the time that we've got, we need to bring about political or social change. And that is the role of proper activists.

I'd like to put forward a case for activists being the people who have done most to shape our world. Yes, the scientists and philosophers have untied all the knots and worked out how to solve our problems, but then our politicians have very often either ignored or abused their inventions or ideas. When it comes to summoning the courage to implement change by revolt through organised or unified protest, to right wrongs and promote reforms for a perceived greater public or environmental good, that's been down to our activists. And there is a long-esteemed column of them lacing back through history.

Think of Spartacus, the gallant former gladiator, and his slave revolts in the first century BC, where 6,000 slaves rebelled and were crucified on the roadside from Capua to Rome by their Roman oppressors. Or one of my heroes, Mahatma Gandhi, who in the 1930s led thousands of protesting Indians on the Salt March as a protest against the oppressive taxes of the government. The blunt response was to imprison 60,000 people, but their suffering eventually fuelled the independence of their nation. Job done. And then there's Martin Luther King Jr, the beautiful Christian minister and activist who became the most visible spokesperson and leader in the civil rights movement from 1955 until he was assassinated in 1968. King's non-violent methods of civil disobedience were inspired by the non-violent activism of Gandhi, and in turn they have inspired the Black Lives Matter movement and Extinction Rebellion.

But for all our admiration of these remarkable people, some seem to harbour a fear of activism, a reaction mischievously stoked by those who wish to resist it, typically because they fear change. Yet activism is prevalent in all our lives as it spans so many activities, from writing letters to newspapers to petitioning elected officials, running or contributing to a political campaign or everyday preferential patronage of businesses or the boycotting of others. Don't we all do at least some of these things? Certainly economic activism is an everyday occurrence for me and many others, whereby we protest by refusing to buy products from a manufacturer whose business model or practices we don't like – that's everything from refusing to pay their taxes in an ethical way to failing to provide proper working conditions to their staff or suppliers. Then there's 'artivism'; don't tell me you don't like or haven't been made to laugh or think by Banksy. In truth of course it's the bigger, noisier activism that gets results more easily, and often 'collective action', where numerous individuals coordinate acts of protest together in order to make a bigger impact. Actions such as these, which are purposeful, organised and often, critically, sustained over a period of time, become known as 'social movements'. Perhaps it's the scale of this mission that scares people – for them there is seemingly a gulf between writing an angry missive to the local paper and heading out on more demonstrative actions like rallies, street marches or participating in strikes, sit-ins or hunger strikes.

SOCIAL JUSTICE = ENVIRONMENTAL JUSTICE

Have you heard the expression recently that environmental justice is social justice? If not, it's definitely something to think about. The term has been used a lot in the last year or so, but especially in light of the Black Lives Matter movement. Justice for the environment and justice for people are two very paramount conversations, but more often than not, they're occurring in separate spaces. We are failing to acknowledge just how intrinsically linked they are. On whichever scale you look at it, globally or locally, people and the planet are so connected that harm to one means parallel harm to the other.

Covid-19 has reminded us of this, as BAME (Black, Asian and minority ethnic) people within the UK are twice as likely to die from the virus, something that has been attributed to the huge social and racial inequality in the UK, and there are similar patterns in respect to climate change. The impacts of a changing environment disproportionately impact people of colour, in both the minority communities in wealthy countries and in poorer developing countries too. According to estimates from the World Bank, climate change may force over 100 million more people into extreme poverty by 2030, which would ironically make them more susceptible to its effects. The World Health Organization predicts that between 2030 and 2050, approximately 250,000 more people could die every year due to increases in malaria, malnutrition and heat stress linked to the climate crisis, mainly in poorer countries throughout Africa and Asia. We already know that people are dying due to the impacts of the climate crisis – to what extent we are not quite sure, but most scientists agree it is likely to be more than we expect. And of course, the brunt of this is already happening in Black and Asian communities. Yet a poll revealed that only 33 per cent of people in the UK think of how climate change impacts different races differently.

The environmental sector has also come under major criticism for lack of diversity and inclusion. Only 3.1 per cent of environmental professionals and 9 per cent of students studying an environmental course identify as non-white. Protest movements, including Extinction Rebellion, have also been under the same scrutiny. There needs to be equal representation of all within the field, because there is strength in our diversity. I am very aware while writing this that I am a white woman who has come from privilege, but I firmly believe that by decolonising the climate movement and by tackling systematic racism, there will be faster and more significant headway in rectifying climate injustices. Patrisse Cullors and Nyeusi Nguvu, members of the Black Lives Matter movement, say, 'Racism is endemic to global inequality.' It would be wrong to say we are all on the same ship when it comes to facing climate change. As it stands, we are all on different ships with varying sizes and defences, just facing the same big tidal wave.

I've fought a lot of battles for wildlife – always distantly motivated by the pain of imagining the cruel deaths of those bulldozed badgers back in the 1980s – and I've lost more than I've won or helped to win. I've just lost another, as my High Court Appeal to call for a Judicial Review into the HS2 rail project was rejected. Taking legal cases for environmental concerns is hard and the success rate is very low. Only 4 per cent are won, an ominous sign of both the parlous state of our environmental legislation and its importance in the eyes of the judiciary. That said, sometimes just taking such cases can lead to progress as they make a lot of noise, raise a lot of interest and motivate others to take other actions. And besides, for me winning isn't ever going to be about crossing a line and being awarded a medal; winning, critically ... is not giving up. And in truth for quite some time I had feared that there'd been a lot of that going on. That was up until April 2019 when my world, and all our worlds, changed ...

In the spring sunshine everything was so colourful, all the banners, flags and badges, all the people. I'm not really a festival sort of bloke, but I approached the exuberant throng happily humming ...

> *'We've heard it all before*
> *We're learning to ignore*
> *You must confess this awful mess*
> *Isn't just a bore'*

I was very conscious that I was unnaturally jovial ...

> *'It's more than we could bear*
> *But you don't really care*
> *Kiss of live to save our life*
> *All you do is stare'*

These are the words of one of my favourite punk songs from way back in 1978 ...

> *'I'm back in full attack*
> *Never give in until they crack*
> *Emergency'*

It's 'Emergency' by 999 and, as I walked towards Oxford Circus, I joyously recalled vocalist Nick Cash cavorting around the stage belting out this song – so, so good! I've always been drawn to the 'never give in until they crack' sentiment, and it was particularly relevant that day, because it was all about the biggest emergency ever. Extinction Rebellion had come to town and were headline news all around the world.

I climbed aboard the pink boat to address the fabulous party and salute their energies and endeavours to put the Climate and Environment Emergency on the map and in everyone's mind. And within a few weeks of their non-violent direct actions that mission had been accomplished – the governments of the UK and many other nations had signed up to the notion that we, the public, were now so aware and concerned about our perilous predicament that we had taken to the streets. And that we would continue to do so. The protests continued throughout the year, a year that saw catastrophic flooding in the Far East and in the north-east of England, devastating wildfires in western America and across huge swathes of Australia, and full frontal belligerent attacks on the environment by a cadre of the worst political leaders in modern times. Finally, Brazil was torched at the behest of its populist president Jair Bolsonaro, one of a number of people from this era that history will not look favourably upon.

But as the rainforests burned it fanned the flames of peaceful revolt around the world. And fast, because in the past activists used pamphlets, flyers and books to disseminate or propagate their messages and attempt to persuade readers of the importance of their cause, but now social media reaches millions instantaneously – it has undoubtedly become both the conservationists' and environmentalists' most important tool. By the end of that April there were 650 XR groups active in 45 different countries.

UNCOOPERATIVE CRUSTIES

Extinction Rebellion certainly hit the headlines in the last few years. Whether you agree with their methods or not, it can't be denied that their actions are initiating important conversations and sparking change. However, I saw on social media that they were called demeaning names and many insinuated that the protesters were just 'tree-hugging hippies'.

During the Autumn Uprising Rebellion of 2019, scientists at Aston University in Birmingham interviewed 300 protesters, fondly known within the movement as rebels, and attended the court hearings of those who were arrested by police for breaching section 14 of the Public Order Act. Their intention was to understand where these rebels had come from, in terms of their backgrounds and beliefs. The results showed that 85 per cent had obtained a university degree, which is over double the national average, and a third had postgraduate qualifications in the form of Masters' or PhDs. Their research also highlighted that XR were very good at engaging new activists, with 10 per cent never having attended a protest before, and half only having been on five or fewer. The XR rebellions have also attracted a much larger age range of people, including families, in comparison to any other environmental movements before it.

You see, when you attend an XR protest it is not about getting arrested and risking your liberty. The majority of rebels are not 'arrestables', but this is something you don't see reflected in the headlines. It's really a festival of art, music, culture and environmental talks all linked by people who share a serious concern for the safety of the environment. Clare Saunders at the University of Exeter suggests that their tactic of blocking roads has been so successful at bringing in more activists because 'the idea of standing in the streets is nowhere near as vulnerable as putting yourself up a tree'.

From my perspective, and that of the many who have joined XR, non-violent direct action is a good way forward so long as it remains peaceful and inclusive. It is perhaps the only way forward unless those with the policy power start to tune in, listen to the scientists and start to realise that the planet's protection is more important than the pounds in their pockets.

After a month the bridges were cleared of protesters and their trees and flowers and the boat was towed away, but the job was done. Of course, it came at a price – policing, loss of income, inconvenience, and a few less well-thought-through actions that made us all cringe. There was limited damage to property – some deliberate, some accidental – but there were no riots, no fights, it was peaceful and thus drew people from all walks of life, young and old, and from all races. Things resumed in the autumn; there was more diversity, more creativity, a range of rebellions had sprung up: XR Farmers, Lawyers for XR, XR Muslims, doctors, families. The appearance of groups like those lawyers, doctors and other respected professionals helped win over more of the sceptics by signalling to wider society that these activists are not all 'un-cooperative crusties', but a broad range of rational people who we take seriously in their professions. Everyone can join in and everyone is needed. 'Affinity groups', as they're known within the study of social movements, have become an increasingly common part of grassroots political organisations in recent years. Bart Cammaerts, a professor of politics and communication at the London School of Economics, says 'We have seen this developing over time in many contemporary movements, where people are valued for their expertise and what they can bring within the larger movement and we see this in other mobilisations, particularly where direct action is being used.' In short it emphasises a competence and breadth of thinking, and with it what some would call 'an air of respectability'.

Another of XR's current successes is that it's a decentralised movement; there is no overall governing structure, and while this undeniably imposes some significant limitations it also grants the freedom for anyone to launch their own XR group. And many have, starting their own 'community groups', which are a way for XR members to connect and work together through communities of shared self-identity, not necessarily of a shared geographical location. Therefore the group can be formed around ethnicity, gender, sexuality, profession, or faith. Many people find joining an XR community group a lot less intimidating than simply showing up to a demonstration in their town or city, and this is certainly the case for those who don't yet think of themselves as typical activists. One of the notable things about XR's recent support is that many of the people who have stepped up and got involved have never protested before. And I can see that some of the demonstrations and protests can look a bit frightening, so that's why having people there who you can already identify with can provide enough reassurance to summon the courage to take to the streets. Plus, if such a group is based around professional skills or a focused interest with accompanying expertise, then even when the banners are back in the cupboard, that collective can continue to correspond and agitate for change.

And change is what we need, most importantly a change of mind. We all need to change our minds, because if you can't change your mind, you will never change the world. So in order to effect that change how many does 'we' need to be? Well, what is essentially reassuring is that it doesn't need to be everyone – after all, with all the time and reason and proof of truth in the world I still can't see Trump, Bolsonaro, Morrison or Johnson signing up to XR. Even better, it doesn't even have to be a majority, as a paper published by Damon Centola from the Annenberg School of Communications, University of Pennsylvania, in *Science* in 2018 outlined. The study involved an online experiment to determine what percentage of total population a minority needs in order to reach the critical mass necessary to reverse a viewpoint the majority holds. The heartening result revealed that the tipping point is just 25 per cent. At that level, the minority arguing their case were able to convince anywhere from 72 to 100 per cent of the population to change their minds. Prior to the campaigning, despite the lobbying efforts of activists, that population had been in 100 per cent agreement about their original position. It's said that power concedes nothing without a demand, that it never has and it never will – but it seems that when a minimum of 25 per cent make that demand the balance of power can be shifted.

However, the same paper also issued a stark warning, 'If you are those people trying to create change, it can be really disheartening', and when even the most committed minorities' efforts start to falter there can be what Centola calls 'a convention to give up'. The trouble, and potential tragedy, is that the struggling minority have no way of knowing when they are at 24.9 per cent and thus on the brink of momentous change. So like all those social change movements that might be approaching that tipping point, it's slow going for XR and there are signs of backsliding, particularly due to the endeavours of the 'billionaire press' and their powerful allies, who are seeking to alienate us from broader society and discredit our imperative demands. And that is frightening, but not as frightening as the rising temperatures of our planet's atmosphere and the losses of its biodiversity.

A total of 1,130 people were arrested during the April 2019 demonstrations in London. But by June just one had been convicted of a public order offence for taking part in the occupations across the city. Demonstrations began again on 6 October and by the 11th over 1,000 more arrests had already been made. Rumours circulated that the then-Home Secretary, Sajid Javid, had put significant pressure on Cressida Dick, the Commissioner of the Metropolitan Police, following the policing of the April uprising. Despite appalling weather thousands supported actions across London and in response on 14 October the police banned all Extinction Rebellion protests from the whole of the capital, starting at 9 p.m. that evening. Later a large number of officers began clearing people and tents from the camp on Trafalgar Square, which until then they had allowed to be a 'safe haven' for protestors. The ban drew instant criticism; even the Mayor of London Sadiq Khan, who would normally be expected to align with the Met, appeared to distance himself from it. The social justice advocacy group Liberty described it as 'a grossly disproportionate move by the Met and an assault on the right to protest'. XR took the matter to court and on 6 November High Court Judges ruled the ban was unlawful, leaving the Met potentially facing claims for false imprisonment from hundreds of wrongfully arrested protesters, something that didn't materialise.

While the ban was an outrageous assault on our democratic right to protest, the behaviour of the police officers on the ground was from my experience exemplary. Faced with the very difficult and novel task of having to arrest polite, smiling and compliant demonstrators, some very elderly, some very young, some as yet unborn – pregnant women – they behaved cordially and gently, with the majority reciprocating those smiles. At every action I spoke at I commended the police and asked for applause, which was always generously forthcoming. One evening I stopped to chat with three officers, all of whom said that if they weren't in their uniforms they'd be standing with us; after it was their world that was burning too.

Megs and I had been doing whatever we could with XR all spring and summer of 2019, and I found myself acting as a spokesperson, championing and validating their brilliant work, rebuffing the criticism and constantly trying to meet their most dedicated campaigners face to face. Megs went on more rallies and worked taking photographs for their media teams. There were many high points for both of us, but perhaps the best was running from

Waterloo to meet a trio of elderly men who had been hunger-striking outside the Conservative Party Headquarters. Trump was in town, security was a bit silly and they had been moved on; nevertheless the police were helpful as ever and we found them. We went just to shake their hands and say 'thank you' because we were in awe of their commitment and determination. They were very welcoming and their smiles will stay with us forever. One day I may well be them; please come and see me.

This was during the Autumn Uprising, and Animal Rebellion had risen. Obviously we went to meet and support them as much as we could, and as usual they and their supporters were all intelligent, focused people, determined to make the world a better place for other species of animal, and therefore our species too. They, with permission, occupied Smithfield's Meat Market. The atmosphere there was great, lots of vegetarian and vegan foods, stands with information about the abuse of farm animals, food science, nutrition – and the thing is they were not all 'ultra-vegans', those intolerant of entertaining a period of transition to a less/no meat diet. We met a range of people with a range of views, all cordial, all engaging. As things progressed it became clear that Animal Rebellion were going to focus on animal welfare, progressing to a meat- and dairy-free diet and all the relevant environmental implications of this, and would maybe not try to shed so much of a light upon the impact of biodiversity loss. This is not a criticism; as I've said before, a man, or a movement's, gotta know its limitations, and addressing the issues they had chosen was a critical and massive task, which we should all support. However, Megs and I felt that biodiversity loss had been overshadowed by XR's tremendous success in garnering global attention for the climate crisis, so during the winter of 2019/2020 we met with all the XR key influencers and mooted the idea of Wildlife Rebellion, a rebellion with a different mission and a different method.

Do you feel happy about being arrested? Some do, many more don't. Megs and I feel our time will come, but we appreciate that a massive number of very informed and concerned people may want to be actively involved in activism, but don't want to go that far ... yet. And thus the XR movement was something they might be shy or mistrustful of, even though they fully understand and support its urgency and overall agenda. Let's cut a short story shorter: we gathered our colleagues, came up with an agenda, refined and defined it, shared it with XR – who liked it and understood it – got some funding, came up with a strategic order of campaigns, met with partners – both corporate and charity, who loved our very obviously non-violent direct actions and objectives – focused on new areas and approaches that would allow entry-level activism and shaped a spring and summer of imaginative actions based around sport, humour, music, art and, just as we were to launch, coronavirus hit and lockdown followed. Both Megs and I, and the team we were assembling, immediately accepted that this was not its time. To launch campaigns that were ultimately designed to promote change in governmental policies – initially nationally and subsequently globally – at a time when governments were about to be tested beyond their capabilities would not be justifiable and appropriate. So, we put it on hold. We cannot say reluctantly, but of course it has been disappointing; the issues we want and need to confront haven't gone away, they have been understandably overshadowed. But, unlike those badgers, crushed and killed all those years ago, we will rise and try to sett the world rights for wildlife. So, when the time is right for us and for you please support Wildlife Rebellion. You'll find us online.

WHAT WILDLIFE?

My grandparents can recall stories from their childhoods detailing a wealth of biodiversity on their doorsteps that I find hard to imagine or relate to. My grandad's voice lifts as he tells me of the cuckoos he would hear from his garden during the 1960s and how the old tatty field and stream behind the house was brimming with grass snakes, tadpoles, frogs and newts. That land in Totton, Southampton, is now a housing estate, and the call of the cuckoo has fallen silent. My nan recalls watching hundreds of butterflies as a little girl along the coast close to where she grew up in Suffolk, but her favourite were the red squirrels in Thetford Forest, which have now disappeared.

You see, for me or other young people, it is really difficult to imagine the abundance of wildlife that was. To get encounters with slowworms or toads today you have to go searching – really searching – but back then, you'd simply have to step outdoors and you were in the wild. I am describing a social phenomenon called shifting baseline syndrome (SBS).

Lizzie Jones at the University of London has been studying SBS, which can be described as the tendency for people to perceive the current environmental climate against very few reference points, so long-term changes are less pronounced.

Essentially, I lack the reference points that my grandparents have and so a nature-depleted country is 'normal' to me. Jones investigated the public's perception of ten different UK birds and asked over 900 people to tell her how abundant they believe the birds to be now versus when they were 18. As may be expected, the results showed that younger people were less likely to recall any abundance changes, which points towards SBS. It makes sense that young people can't remember how much wildlife there once was and therefore can't draw those comparisons, but it poses huge problems for contemporary conservation. And regardless of age, we all fall for it. Hearing my grandad's stories from the sixties I can't help but think 'wow', but what was it like for his grandfather, and his grandfather before that? You see, we strive to make things better within our own baselines and once we think we have succeeded we stop trying. Based on my experiences, our wildflower patch saw so many peacock butterflies this year; it seemed like loads – which is true for 2020, but would not be the case for those people making similar observations in the sixties, fifties, forties, etc.

There is always something more we can do to improve, because the scope of the issue is much bigger than us and the limitations of our memories. Young people may not be able to remember these changes, yet they are so well informed and still actively and peacefully fight for their environment. We have listened to our grandparents' and parents' stories, we have read the books and science and we've seen the photos – we are fighting for a fairytale. A fairytale of cuckoos, butterflies and red squirrels.

And now, the end is near, and so we face the final sentence. So my friends, I'll say it clear, I'll state my case, of which I'm so, so very certain ... In the past few years I've sensed a chronic acceptance of our beautiful nature vanishing. It manifests itself in so many conversations we have where we share those shocking statistics of each decline or destruction among ourselves, like a vicious game of top trumps, seemingly to the extent that those horrific numbers have utterly lost their meaning. What we seem to have forgotten is that they are a death toll, that those figures are a numerical transcript of the dwindling voices of vanished millions, that they sound a tragic echo of a recent time of far more plentiful life. I think it's time to all give ourselves a hard slap. We must wake up from our complacent stupor, because as every one of us knows we are presiding over a very real ecological apocalypse and conducting a mass extermination in our own backyard, literally my backyard and yours. Our gardens, our communities, our countryside; the places we love most of all. But critically, it is not too late; there is hope and a lot more besides.

In 2018 I conducted a UK bioblitz and with the help of 785 recorders and thirteen recording centres our team clocked up a notable 4828 different species in ten days. Loads of exciting plants, animals and fungi, but perhaps even more impressively and importantly lots of passionate, energetic, skilful, imaginative and creative conservationists. Some were in gardens, some in community wildlife areas, others on wildlife-friendly farms or big flashy nature reserves; all were making a difference in their own important and impressive ways. Those people were you, or people like you. And now in the wake of the coronavirus lockdown even more of us have connected with, and found a love for and a value in nature. In even the humblest, everyday, still-common creatures that have lifted our spirits in the darkest days of a terrible crisis. And when it comes to protecting them, we have plenty of tools in the conservation box; we can rebuild, restore, reinstate or reintroduce. We can march, lobby, sign petitions, we can demonstrate. We can take action. We can make a difference.

So all you farmers, foresters, reserve wardens, teachers, students and children, all of you 'ologists', scientists, artists, writers and bloggers, you activists, volunteers, gardeners, all of you who have taken a bit of solace and respite from the horrors of 2020 by finding, engaging with and loving nature, can you please see that this is not your last chance to make a difference, it is our species' last chance to make that difference. We don't all have to agree about all the details, but we must agree on our shared agenda. We must stand shoulder to shoulder with all who care enough to take some action and be part of making a difference. You – please, you – need to make that difference.

Our wildlife needs us, and it needs you more than ever.

ACKNOWLEDGEMENTS

We would like to thank Fabian Harrison for all his exceptional assistance with all our campaigning and for his unparalleled technical expertise, which allowed the Self Isolating Bird Club to reach as many people as it did and does. Cate Crocker has performed miracles behind the scenes to organise all the broadcasts and correspond with all the contributors as well as manage the social media. Thanks must also go to Lucy Groves (Project Officer, The White Stork Project/Durrell Wildlife Conservation Trust), Ian Thomson (Head of Investigations, RSPB Scotland), Dr Zoe Randle and Dr Caroline Bulman (Butterfly Conservation), David Ramsden (Head of Conservation, The Barn Owl Trust), Isabella Tree (Knepp Wildland), Jeremy Roberts (Programme Manager, Cairngorms Connect), Bob Elliot (Director, OneKind), Dr Ruth Tingay (Raptor Persecution UK), Roy Dennis (The Roy Dennis Wildlife Foundation), Kevin Cumming (Project Leader, Langholm Moor Community) and Martin Lines (Chairman, Nature Friendly Farming Network UK). All were kind enough to read over relevant sections of the text and pass comment and critical advice. Both Iolo Williams and Dara McAnulty wrote authoritatively about the governments' statutory agencies in Wales and Northern Ireland respectively; thank you for your insights.

We would also like to thank Kate Hewson at Two Roads for her patience and sterling work as our editor, and the rest of the Two Roads team, Rosie Gailer, Yassine Belkacemi, Jess Kim and Emma Petfield, for their commitment and hard work on this project. David Foster, our agent and friend, took meticulous care of the business allowing us to focus on writing, and thanks to Sid and Nancy for necessary distraction, barking, face-licking and mess-making. Charlotte made us breakfast, lunch and dinner when we were too busy typing to breathe . . . and they were all tasty too. And finally, thanks to all the followers of the Self Isolating Bird Club for your inspirational enthusiasm, knowledge and passion through 'lockdown' and beyond – it's been a joy being part of your community.

ABOUT THE AUTHORS

Chris Packham is one of the UK's leading naturalists and wildlife TV presenters, inspiring audiences young and old to take notice of, get involved with and care for our natural environment. He is currently one of the hosts of BBC 2's *Springwatch*, *Autumnwatch* and *Winterwatch*. His autobiography, *Fingers in the Sparkle Jar* was a number one *Sunday Times* bestseller, and a Radio 4 Book of the Week.

As a campaigner Chris is a vociferous opponent of the badger cull, HS2 and has called for the banning of driven grouse shooting. Along with Mark Avery and Ruth Tingay he is a founder of Wild Justice, an entity that seeks legal reforms to protect the UK's wildlife. In 2018 he organised the 'People's Walk for Wildlife' and collated 'A People's Manifesto for Wildlife'. He was awarded a CBE in 2019 for services to conservation.

Megan McCubbin is a zoologist, wildlife TV presenter, conservationist and environmental activist. Her interest stems from a childhood growing up around wildlife and from time spent at The Wildheart Trust on the Isle of Wight, which specialises in the rescue and rehabilitation of ex-circus animals. She most recently presented *Springwatch* and *Autumnwatch*, and is a co-founder of Wildlife Rebellion and The Self-Isolating Bird Club on social media.

She is a youth ambassador for the League Against Cruel Sports, is campaigning to end captive elephant abuse and is a keen and prize winning photographer. In 2019 she was appointed as the coordinator and judge of Young Bird Photographer of the Year.